REA's Books Are

They have rescued lots of grades and more!

(a sample of the <u>hundreds of letters</u> REA receives each year)

"Your books are great! They are very helpful, and have upped my grade in every class. Thank you for such a great product."

Student, Seattle, WA

"Your book has really helped me sharpen my skills and improve my weak areas. Definitely will buy more."

Student, Buffalo, NY

"Compared to the other books that my fellow students had, your book was the most useful in helping me get a great score."

Student, North Hollywood, CA

"I really appreciate the help from your excellent book. Please keep up your great work."

Student, Albuquerque, NM

"Your book was such a better value and was so much more complete than anything your competition has produced (and I have them all)!"

Teacher, Virginia Beach, VA

(more on next page)

PRE-CALCULUS

By Ernest Woodward, Ed.D.

Professor of Mathematics
Austin Peay State University, Clarksville, Tenn.

**and the Staff of
Research & Education Association**

Research & Education Association
61 Ethel Road West
Piscataway, New Jersey 08854

SUPER REVIEW®
OF PRE-CALCULUS

Year 2005 Printing

Printed in the United States of America

Library of Congress Control Number 00-130288

International Standard Book Number 0-87891-088-3

What this **Super Review**® Will Do for You

REA's **Super Review** provides **all you need to know** to excel in class and succeed on midterms, finals, and even pop quizzes.

Think of this book as giving you access to your own private tutor. Here, right at your fingertips, is a brisk review to help you not only understand your textbook but also pick up where even some of the best lectures leave off.

Outstanding **Super Review** features include...

- Comprehensive yet concise coverage
- Targeted preparation for subject tests
- Easy-to-follow **Q** & **A** format that helps you master the subject matter
- End-of-chapter quizzes that provide pretest tune-up

We think you'll agree that, whether you're prepping for your next test or want to be a stronger contributor in class, **REA's Super Review** truly provides **all you need to know!**

Larry B. Kling
Super Review Program Director

CONTENTS

CHAPTER 1

Sets, Numbers, Operations, and Properties

1.1 Sets

A collection of objects categorized together is called a set. There are two standard ways to represent sets, the roster method and the set-builder method. For example,

$$\{2, 3, 4\}$$

is in roster form and

$$\{x \mid x \text{ is a counting number between 1 and 5}\}$$

is in set-builder form, although both describe the same set.

The symbol "$\in$" is used to represent "is an element of" and the symbol "$\subseteq$" is used to represent "is a subset of." Here are two important definitions concerning sets:

$A \subseteq B$ if and only if every element of A is an element of B.

$A = B$ if and only if $A \subseteq B$ and $B \subseteq A$.

Two important set operations are union, denoted by "∪," and intersection, denoted by "∩," defined below.

$$A \cup B = \{ x \mid x \in A \text{ or } x \in B \}$$

$$A \cap B = \{ x \mid x \in A \text{ and } x \in B \}$$

Problem Solving Example:

If $A = \{2, 3, 5, 7\}$ and $B = \{1, -2, 3, 4, -5, 6\}$, find (a) $A \cup B$ and (b) $A \cap B$.

(a) $A \cup B$ is the set of all elements in A or in B or in both A and B, with no element included twice in the union set.

$$A \cup B = \{1, 2, -2, 3, 4, 5, -5, 6, 7\}$$

(b) $A \cap B$ is the set of all elements in both A and B.

$$A \cap B = \{3\}$$

Sometimes two sets have no elements in common. Let $S = \{3, 4, 7\}$ and $T = \{2, -4, 6\}$. What is the intersection of S and T? In this case $S \cap T$ has no elements. Hence $S \cap T = \emptyset$, empty set. In that case, the sets are said to be disjoint.

The set of all elements entering a discussion is called the universal set, U, When the universal set is not given, we assume it to be the set of real numbers. The set of all elements in the universal set that are not elements of A is called the complement of A, written $\overline{A}$.

1.2   Real Numbers and their Components

Real numbers provide the basis for most pre-calculus mathematics topics. The set of all real numbers has various components. These components are the set of all natural numbers, N, the set of all whole numbers, W, the set of all integers, I, the set of all rational numbers, Q, and the set of all irrational numbers, S. Then,

$$N = \{1, 2, 3, \ldots\},$$

$$W = \{0, 1, 2, 3, \ldots\},$$

$I = \{\dots, -3, -2, -1, 0, 1, 2, 3, \dots\}$,

$Q = \{a/b \mid a, b \in I \text{ and } b \neq 0\}$,

and $S = \{x \mid x$ has a decimal name which is nonterminating and does not have a repeating block$\}$.

It is obvious that $N \subseteq W$, $W \subseteq I$, and $I \subseteq Q$, but a similar relationship does not hold between Q and S. More specifically, the decimal names for elements of Q are

(1) terminating or

(2) nonterminating with a repeating block.

For example, $\frac{1}{2} = .5$ and $\frac{1}{3} = .333\dots$. This means that Q and S have no common elements. Examples of irrational numbers include .101001000..., π, and $\sqrt{2}$.

All real numbers are normally represented by R, and $R = Q \cup S$. This means that every real number is either rational or irrational. A nice way to visualize real numbers geometrically is to see that real numbers can be put in a one-to-one correspondence with the set of all points on a line.

Problem Solving Example:

Classify each of the following numbers into as many different sets as possible. Example: real, integer, rational ....

(1) 0

(2) 9

(3) $\sqrt{6}$

(4) $\frac{1}{2}$

(5) $\frac{2}{3}$

(6) 1.5

(1)   Zero is a real, whole, rational number, and an integer.

(2)   9 is a real, whole, rational, natural number, and an integer.

(3) $\sqrt{6}$ is an irrational, real number.

(4) $\frac{1}{2}$ is a rational, real number.

(5) $\frac{2}{3}$ is a rational, real number.

(6)   1.5 is a rational, real number.

1.3   Real Number Properties of Equality

The standard properties of equality involving real numbers are:

Reflexive Property of Equality

For each real number a,

$a = a$.

Symmetric Property of Equality

For each real number a, for each real number b,

if $a = b$, then $b = a$.

Transitive Property of Equality

For each real number a, for each real number b, for each real number c,

if $a = b$ and $b = c$, then $a = c$.

Other properties of equality are listed in Chapter 4.

Problem Solving Example:

If $a = 1$ and $c = 1$, by the transitive property of equality, how are a and c related?

First, the transitive property states that if $a = b$ and $b = c$, then $a = c$.

In this case $a = 1$, and $c = 1$ or $1 = c$. Therefore, let $b = 1$ and thus $a = 1 = c$. Therefore a and c are equal.

1.4 Real Number Operations and Their Properties

The operations of addition and multiplication are of particular importance. As a result, many properties concerning those operations have been determined and named. Here is a list of the most important of these properties:

Closure Property of Addition

For every real number a, for every real number b,

$a + b$

is a real number.

Closure Property of Multiplication

For every real number a, for every real number b,

ab

is a real number.

Commutative Property of Addition

For every real number a, for every real number b,

$a + b = b + a.$

Commutative Property of Multiplication

For every real number a, for every real number b,

$ab = ba.$

Associative Property of Addition

For every real number a, for every real number b, for every real number c,

$(a + b) + c = a + (b + c).$

Associative Property of Multiplication

For every real number a, for every real number b, for every real number c,

$(ab)c = a(bc).$

Identity Property of Addition

For every real number a,

$a + 0 = 0 + a = a.$

Identity Property of Multiplication

For every real number a,

$a \times 1 = 1 \times a = a.$

Inverse Property of Addition

For every real number a, there is a real number $-a$ such that

$a + -a = -a + a = 0.$

Inverse Property of Multiplication

For every real number a, $a \neq 0$, there is a real number a^{-1} such that

$a \times a^{-1} = a^{-1} \times a = 1.$

Distributive Property

For every real number a, for every real number b, for every real number c,

$$a(b + c) = ab + ac.$$

The operations of subtraction and division are also important, but less important than addition and multiplication. Here are the definitions for these operations.

For every real number a, for every real number b, for every real number c,

$$a - b = c \text{ if and only if } b + c = a.$$

For every real number a, for every real number b, for every real number c,

$$a \div b = c \text{ if and only if } c \text{ is the unique real number such that}$$
$bc = a$.

The definition of division eliminates division *by* 0. Thus, for example, $4 \div 0$ is undefined, $0 \div 0$ is undefined, but $0 \div 4 = 0$.

In many instances, it is possible to perform subtraction by first converting a subtraction statement to an addition statement. This is illustrated below.

For every real number a, for every real number b,

$$a - b = a + (-b).$$

In a similar way, every division statement can be converted to a multiplication statement. Use the following model:

For every real number a, for every real number b, $b \neq 0$,

$$a \div b = a \times b^{-1}.$$

Problem Solving Examples:

 Evaluate $2 - \{5 + (2 - 3) + [2 - (3-4)]\}$

 When working with a group of nested parentheses, we evaluate the innermost parenthesis first.
Thus,

$$= 2 - \{5 + (2 - 3) + [2 - (-1)]\}$$
$$= 2 - \{5 + (-1) + [2 + 1]\}$$
$$= 2 - \{5 + (-1) + 3\}$$
$$= 2 - \{4 + 3\}$$
$$= -5.$$

 Simplify $x = a + 2[b - (c - a + 3b)]$.

 When working with several groupings, we perform the operations in the innermost parenthesis first, and work outward.
Thus, we first subtract $(c - a + 3b)$ from b:

$$x = a + 2[b - (c - a + 3b)] = a + 2(b - c + a - 3b)$$

Combining terms,

$$= a + 2(-c + a - 2b)$$

distributing the 2,
$$= a - 2c + 2a - 4b$$

combining terms,
$$= 3a - 2c - 4b$$

To check that $a + 2[b - (c - a + 3b)]$ is equivalent to $3a - 2c - 4b$, replace a, b, and c by any values. Letting $a = 1$, $b = 2$, $c = 3$, the original form $a + 2[b - (c - a + 3b)] = 1 + 2[2 - (3 - 1 + 3 \times 2)]$

$$= 1 + 2[2 - (3 - 1 + 6)]$$

$$= 1 + 2[2 - 8]$$
$$= 1 + 2(-6)$$
$$= 1 + (-12)$$
$$= -11$$

The final form, $3a - 2c - 4b$

$$= 3(1) - 2(3) - 4(2)$$
$$= 3 - 6 - 8$$
$$= -11$$

Thus, both forms yield the same result.

1.5 Complex Numbers

As indicated above, real numbers provide the basis for most pre-calculus mathematics topics. However, on occasion there arise situations in which real numbers by themselves are not enough to explain what is happening. As a result, complex numbers developed.

Returning momentarily to real numbers, the square of a real number cannot be negative. More specifically, the square of a positive real number is positive, the square of a negative real number is positive, and the square of 0 is 0. Then i is defined to be a number with a property that

$$i^2 = -1.$$

Obviously, i is not a real number. C is then used to represent the set of all complex numbers and

$$C = \{ a + bi \mid a \text{ and } b \text{ are real numbers } \}.$$

Here are the definitions of addition, subtraction, and multiplication of complex numbers.

Suppose $x + yi$ and $z + wi$ are complex numbers. Then

$$(x + yi) + (z + wi) = (x + z) + (y + w)i$$

$$(x + yi) - (z + wi) = (x - z) + (y - w)i$$
$$(x + yi) \times (z + wi) = (xz - yw) + (xw + yz)i.$$

Division of two complex numbers is usually accomplished with a special procedure that involves the conjugate of a complex number. The conjugate of $a + bi$ is denoted by:

$$\overline{a + bi} \quad \text{and} \quad \overline{a + bi} = a - bi.$$

Also, $\quad (a + bi)(a - bi) = a^2 + b^2.$

The usual procedure for division is illustrated below.

$$\frac{x + yi}{z + wi} = \frac{x + yi}{z + wi} \times \frac{z - wi}{z - wi}$$

$$= \frac{(xz + yw) + (-xw + yz)i}{z^2 + w^2}$$

$$= \frac{xz + yw}{z^2 + w^2} + \frac{-xw + yz}{z^2 + w^2}i$$

All the properties of real numbers described in the previous section carry over to complex numbers; however, those properties will not be stated again.

If a is a real number, then a can be expressed in the form $a = a + 0i$. Hence, every real number is a complex number, and $R \subseteq C$.

Problem Solving Examples:

 Write each of the following in the form $a + bi$.

(a) $(2 + 4i) + (3 + i)$

(b) $(2 + i) - (4 - 2i)$

(c) $(4 - i) - (6 - 2i)$

(d) $3 - (4 + 2i)$

A

(a) $(2+4i)+(3+i)$ $= 2+4i+3+i$

$= (2+3)+(4i+i)$

$= 5+5i$

(b) $(2+i)-(4-2i)$ $= 2+i-4+2i$

$= (2-4)+(i+2i)$

$= -2+3i$

(c) $(4-i)-(6-2i)$ $= 4-i-6+2i$

$= (4-6)+(-i+2i)$

$= -2+i$

(d) $3-(4+2i)$ $= 3-4-2i$

$= (3-4)-2i$

$= -1-2i$

 Q What is the conjugate of $3-2i$ and the conjugate of $5+7i$?

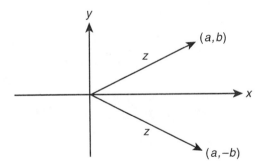

A Any complex number may be interpreted as an ordered pair in the plane with the real component designated by the x value and the imaginary part designated by the y value. The conjugate of a complex number is that number which when multiplied by the original complex number yields a product which is purely real.

Geometrically, the complex conjugate is a reflection of the complex number through the x-axis. The complex conjugate of $3 - 2i$ is $3 + 2i$, $(3 - 2i)(3 + 2i) = 13$.

The conjugate of $5 + 7i$ is $5 - 7i$, $(5 + 7i)(5 - 7i) = 74$.

The conjugate of a pure real number a, which can be written $a + 0i$, is merely itself or $a - 0i$. Geometrically we see that the reflection of a real number is actually itself. The conjugate of a pure imaginary number bi is $-bi$. The conjugate of a complex number $a + bi$ is $a - bi$.

Find the product $(2 + 3i)(-2 - 5i)$.

Using the following method: product of first elements + product of outer elements + product of inner elements + product of last elements:

$$(2 + 3i)(-2 - 5i) = 2(-2) + 2(-5i) + 3i(-2) + 3i(-5i)$$
$$= -4 - 10i - 6i - 15i^2$$
$$= -4 - 16i - 15i^2$$

Recall $i^2 = -1$, hence, $= -4 - 16i - 15(-1)$
$$= -4 - 16i + 15$$
$$= 11 - 16i$$

The same result is obtained by using the distributive law.
$$(2 + 3i)(-2 - 5i) = (2 + 3i)(-2) - (2 + 3i)5i$$
$$= -4 - 6i - 10i - 15i^2 = 11 - 16i.$$

In other words, if one multiplies $2 + 3i$ and $-2 - 5i$ as if they were polynomials and replaces i^2 by -1, then the correct product is obtained.

Simplify $\dfrac{3-5i}{2+3i}$.

To simplify $\dfrac{3-5i}{2+3i}$ means to write the fraction without an imaginary number in the denominator. To achieve this, we multiply the fraction by another fraction which is equivalent to unity (so that the value of the original fraction is unchanged), which will transform the expression in the denominator to a real number. A fraction with this property must have the complex conjugate of the expression in the denominator of the original fraction as its numerator and denominator. The complex conjugate must be chosen because of its special property that when multiplied by the original complex number the result is real.

Note: $a + bi$; its complex conjugate is $a - bi$ or they can be said to be conjugates of each other. To multiply, notice that $(a+bi)(a-bi)$ is the factored form of the difference of two squares. Thus we obtain

$$(a)^2 - (bi)^2; i^2 = -1; \quad (a)^2 - (-1)(b)^2 \quad \text{or} \quad a^2 + b^2.$$

$$\frac{3-5i}{2+3i} \times \frac{2-3i}{2-3i} = \frac{6-9i-10i+15i^2}{4-9i^2}$$

$$= \frac{6-19i-15}{4+9}$$

$$= \frac{-9-19i}{13} \quad \text{or} \quad \frac{-9}{13} - \frac{19}{13}i.$$

Since the resulting fraction has a rational number in the denominator, we have rationalized the denominator.

Quiz: Sets, Numbers, Operations, and Properties

1. Let $U = \{1,2,3,4,5,6,7,8,9,10\}$, $A = \{2,4,6\}$, $B = \{1,4,8,10\}$. Which of the following is the set $\overline{(A \cap B)}$?

    (A) $\{1,2,3,5,6,7,8,9,10\}$ (D) Ø

    (B) $\{4\}$ (E) $\{1,2,4,6,8,10\}$

    (C) $\{3,5,7,9\}$

2. Let $n(A)$ denote the number of elements in set A. If $n(A) = 10$, $n(B) = 12$, and $n(A \cap B) = 3$, how many elements does $A \cup B$ contain?

    (A) 10   (D) 19

    (B) 12   (E) 22

    (C) 15

3. If $A \subseteq C$ and $B \subseteq C$, which of the following statements is true?

    (A) The set $A \cup B$ is also a subset of C.

    (B) The complement of A is also a subset of C.

    (C) The complement of B is also a subset of C.

    (D) The union of $\overline{A}$ and $\overline{B}$ contains C.

    (E) C is the universal set.

4. If $\overline{A} = 2 + 3i$ and $\overline{B} = -1 + 2i$, then $\overline{A} + \overline{B} =$

    (A) $2 + 5i$.   (D) $3 + 2i$.

    (B) $1 + 5i$.   (E) $3 + 5i$.

    (C) $3 + i$.

5. If $a = 3 + 2i$ and $b = 1 + 4i$ and $i = \sqrt{-1}$, then $ab =$

    (A) $4 + 6i$.   (D) $-6 + 13i$.

    (B) $1 + 2i$.   (E) $6 + 14i$.

    (C) $-5 + 14i$.

6. If $i^2 = -1$ and $x = 2 + 5i$, then $\dfrac{1}{x} =$

    (A) $\dfrac{2}{27} + \dfrac{5}{27}i$.   (D) $2 + 5i$.

    (B) $\dfrac{2}{29} - \dfrac{5}{29}i$.   (E) $3 + 10i$.

    (C) $\dfrac{1}{2} + \dfrac{1}{5}i$.

7. If we write the expression

$$\frac{1}{1-i} - \frac{1}{i}$$

in the form $a + bi$ ($i^2 = -1$), the result will be

    (A) $3 + 2i$.   (D) $\dfrac{1}{2} + \dfrac{3}{2}i$.

    (B) $\dfrac{5}{3} - \dfrac{1}{3}i$.   (E) $\dfrac{1}{2} + 2i$.

    (C) $\dfrac{1}{2} + 5i$.

8. What is $(3+2i)(3-2i)$?

    (A) $5+12i$.   (D) 5.

    (B) 13.   (E) 6.

    (C) $5-12i$.

9. What is the multiplicative inverse of $\dfrac{1}{n+1}$?

    (A) $n-1$.   (D) $\dfrac{1}{n}$.

    (B) n.   (E) $1n-1$.

    (C) $n+1$.

10. What is the additive inverse of -3?

    (A) 3.   (D) 0.

    (B) $\dfrac{1}{3}$.   (E) -3.

    (C) $-\dfrac{1}{3}$.

ANSWER KEY

1.	(A)	6.	(B)
2.	(D)	7.	(D)
3.	(A)	8.	(B)
4.	(B)	9.	(C)
5.	(C)	10.	(A)

CHAPTER 2

Coordinate Geometry

2.1 Real Numbers and the Coordinate Line

As was mentioned in Section 1.2, real numbers can be put in a one-to-one correspondence with the points on a line. The common practice is to select a point, call it the origin, and assign to it the number 0. Then the positive real numbers are identified with the points to the right of the origin, and the negative real numbers are associated with the points to the left of the origin.

Figure 2.1

If P_1 and P_2 are associated with the real numbers x_1 and x_2, then the distance (d) between P_1 and P_2 is defined to be

$d = |x_2 - x_1|$. (Absolute value is defined in Section 4.3.)

There is an ordering on the real numbers analogous to the ordering of points on the line.

$a < b$ indicates that a is less than b. The point corresponding to the real number b lies to the right of the point corresponding to the real number a.

$a \leq b$ indicates that $a < b$ or $a = b$.

An uninterrupted portion of a number line is called an interval. The symbol (a, b) is called an open interval and is defined below.

$(a, b) = \{\ x \mid x \text{ is a real number and } a < x < b\}$

The graph of this interval is shown below.

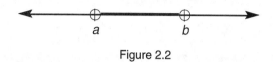

Figure 2.2

In this graph, the open circles at the two ends indicate that the points associated with a and b are not included in the interval.

The symbol $[a, b]$ is called a closed interval, and its definition and graph follow.

$[a, b] = \{\ x \mid x \text{ is a real number and } a \le x \le b\}$

Figure 2.3

In this instance, the closed dots indicate that the points associated with a and b are included in the interval. Similarly, $(a, b]$ and $[a, b)$ are called half open intervals, and

$(a, b] = \{\ x \mid x \text{ is a real number and } a < x \le b\}$

$[a, b) = \{x \mid x \text{ is a real number and } a \le x < b\}.$

2.2 Cartesian Coordinates

In a coordinate system, a specific horizontal line is called the x-axis, and a specific vertical line is called the y-axis. The point of intersection of these lines is called the origin. (See Figure 2.4.)

By starting at the origin, one way to get to point P is to move two units to the left and then move one unit up. Then P is associated with $(-2, 1)$, and $(-2, 1)$ are called the coordinates of P. More specifically, -2 is the x-coordinate of P and 1 is the y-coordinate of P.

Lines and Segments

The axes divide the plane into four quadrants which are labeled I, II, III, and IV, as indicated in the illustration below. It should be noted that (a, b) has one meaning in Cartesian coordinates and another distinctly different meaning in the coordinate geometry of a line.

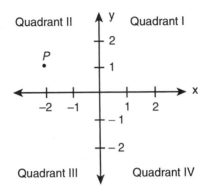

Figure 2.4

The notation $P(x, y)$ means that P is a point whose coordinates are (x, y).

2.3   Lines and Segments

Using the Pythagorean theorem, it is fairly easy to establish that the distance between $P(x_1, y_1)$ and $Q(x_2, y_2)$ is:

$$d = \sqrt{\left(x_2 - x_1\right)^2 + \left(y_2 - y_1\right)^2}.$$

Also, the coordinates of the midpoint of the segment joining $P(x_1, y_1)$ and $Q(x_2, y_2)$ are:

$$\left(\frac{x_1 + x_2}{2}, \frac{y_1 + y_2}{2}\right).$$

The slope of a line is a number which indicates the slant, or direction, of the line. The symbol "m" is used to represent the slope of a line. For a nonvertical line which passes through $P(x_1, y_1)$ and $Q(x_2, y_2)$, the slope is:

$$m = \frac{y_2 - y_1}{x_2 - x_1}.$$

From properties of similar triangles, it is easy to show that the slope of a nonvertical line is unique and its calculation is independent of the two points selected. It is somewhat common to use the symbol Δy to represent $y_2 - y_1$ and the symbol Δx to represent $x_2 - x_1$. Using these symbols,

$$m = \frac{\Delta y}{\Delta x}$$

A good intuitive interpretation of slope is very important in calculus and precalculus mathematics. Lines with various slopes are pictured below.

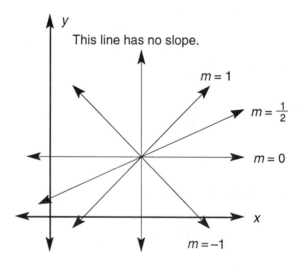

Figure 2.5

It can be determined whether or not two lines are parallel or perpendicular by inspecting their slopes. Here is how this can be done when lines l_1 and l_2 have slopes m_1 and m_2.

$l_1 \parallel l_2$ if and only if $m_1 = m_2$

$l_1 \perp l_2$ if and only if $m_1 \times m_2 = -1$

Since any two vertical lines are parallel, and since a slope does not exist for a vertical line, when two lines do not have a slope they are parallel. Also, since any vertical line is perpendicular to any horizontal line, any line with slope 0 is perpendicular to any line which does not have a slope.

Suppose points $P(0, b)$ and $Q(a, 0)$ are points on a line. Then, a is called the x-intercept of the line and b is called the y-intercept of the line. An equation of the form

$$\frac{x}{a} + \frac{y}{b} = 1$$

is said to be in intercept form. Here is the graph of this line.

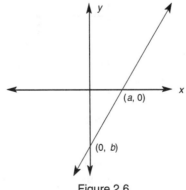

Figure 2.6

An equation of the form

$$y = mx + b$$

is said to be in the slope-intercept form. Here is a graph of this line.

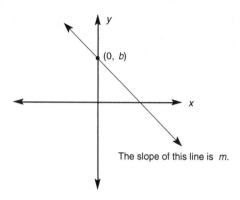

The slope of this line is m.

Figure 2.7

If (x_1, y_1) is a point on a line, then an equation of the form

$$y - y_1 = m(x - x_1)$$

is said to be in point-slope form. Here is a graph of this line.

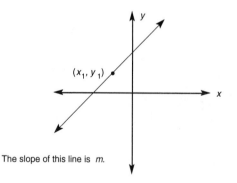

The slope of this line is m.

Figure 2.8

It is somewhat common for someone to mention *the* equation of a line. This is, of course, incorrect. A line has many equations but the ones mentioned above are important enough to be given specific names.

The distance between a point and a line refers to the shortest possible distance between the point and the line. This is, of course, the perpendicular distance between the point and the line. For a point $P(x_0, y_0)$ and a line with slope m and y-intercept b, this distance is given by:

Lines and Segments

$$d = \frac{|mx_0 + b - y_0|}{\sqrt{1 + m^2}}.$$

Problem Solving Examples:

Find the distance between the given pair of points, and find the slope of the line segment joining them.

(3, -5), (2, 4)

Let (3, -5) be $P_1(x_1, y_1)$, and let (2, 4), be $P_2(x_2, y_2)$. By the distance formula,

$$d = \sqrt{(x_2 - x_1)^2 + (y_2 - y_1)^2}, \quad \text{the}$$

the distance between the points (3, -5) and (2, 4) is:

$$d = \sqrt{(2-3)^2 + (4-(-5))^2}$$
$$= \sqrt{(-1)^2 + (4+5)^2}$$
$$= \sqrt{1 + (9)^2}$$
$$= \sqrt{1 + 81}$$
$$= \sqrt{82}$$

The slope of the line joining the points (3, -5) and (2, 4) is given by the formula:

$$\text{slope} = m = \frac{y_2 - y_1}{x_2 - x_1}$$

Again, let (3, -5) be $P_1(x_1, y_1)$ and let (2, 4) be $P_2(x_2, y_2)$. Then the slope is:

$$m = \frac{4-(-5)}{2-3}$$

$$= \frac{4+5}{-1}$$

$$= \frac{9}{-1}$$

$$= -9$$

 Find the midpoint of the segment from $R(-3, 5)$ to $S(2, -8)$.

 The midpoint of a line segment from (x_1, y_1) to (x_2, y_2) is given by

$$\left(\frac{x_1 + x_2}{2}, \frac{y_1 + y_2}{2} \right)$$

the abscissa being one half the sum of the abscissas of the endpoints and the ordinate one half the sum of the ordinates of the endpoints. Let the coordinates of the midpoint be $P(x_0, y_0)$. Then,

$$x_0 = \frac{1}{2}(-3+2) = -\frac{1}{2} \qquad y_0 = \frac{1}{2}[5+(-8)] = \frac{1}{2}(-3) = -\frac{3}{2}$$

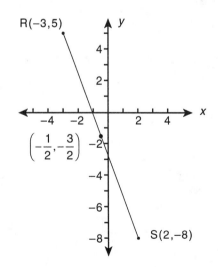

Thus, the midpoint is $P\left(-\dfrac{1}{2}, -\dfrac{3}{2}\right)$.

 If $f(x) = -2x - 5$, find the
(a) slope,
(b) x-intercept,
(c) y-intercept,
(d) Graph the function

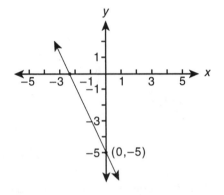

A $f(x) = mx + b$ is called a linear function where m and b are constants. m is the slope of the line, and b is the y-intercept of the line. In this case, $f(x) = -2x - 5$,

(a) slope: $m = -2$

(b) The x-intercept is located on the x-axis where $f(x) = 0$. Then we solve for x.

Here, $0 = -2x - 5$

Solving for x, $x = \dfrac{-5}{2}$

(c) y-intercept: $b = -5$

(d) The graph is shown above.

Show that the points $A(-2, 4)$, $B(-3, -8)$, and $C(2, 2)$ are vertices of a right triangle.

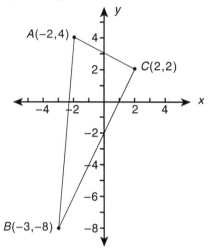

If triangle ABC is a right triangle, then $a^2 + b^2 = c^2$; that is, the sum of the squares of the legs equals the square of the hypotenuse by the Pythagorean Theorem. It is also true that if the sides are in this relationship, then the triangle is right.

Thus we compute the distance from B to C which is side a, the distance from C to A which is side b, and the distance from A to B which is side c.

The formula for the distance between two points (x_1, y_1) and (x_2, y_2) is $d = \sqrt{(x_2 - x_1)^2 + (y_2 - y_1)^2}$

Thus the distance from B to C, from $(-3, -8)$ to $(2, 2)$, is

$$\sqrt{[2 - (-3)]^2 + [2 - (-8)]^2} = \sqrt{(2+3)^2 + (2+8)^2}$$
$$= \sqrt{5^2 + 10^2}$$
$$= \sqrt{25 + 100}$$
$$= \sqrt{125}$$

Hence side $a = \sqrt{125}$.

The distance from C to A, from $(2, 2)$ to $(-2, 4)$, is

$$\sqrt{(-2-2)^2 + (4-2)^2} = \sqrt{(-4)^2 + 2^2}$$
$$= \sqrt{16+4}$$
$$= \sqrt{20}$$

Hence side $b = \sqrt{20}$.

The distance from A to B, from $(-2, 4)$ to $(-3, -8)$, is

$$\sqrt{[-3-(-2)]^2 + (-8-4)^2} = \sqrt{(-3+2)^2 + (-12)^2}$$
$$= \sqrt{(-1)^2 + (12)^2}$$
$$= \sqrt{1+144}$$
$$= \sqrt{145}$$

Hence side $c = \sqrt{145}$.

Now, if triangle ABC is a right triangle, $a^2 + b^2 = c^2$.

Replacing, a by $\sqrt{125}$, b by $\sqrt{20}$, and c by $\sqrt{145}$,

we obtain,

$$\left(\sqrt{125}\right)^2 + \left(\sqrt{20}\right)^2 = \left(\sqrt{145}\right)^2$$

since $\left(\sqrt{a}\right)^2 = \sqrt{a} \times \sqrt{a} = \sqrt{a^2} = a$,

$$\left(\sqrt{125}\right)^2 = 125,$$

$$\left(\sqrt{20}\right)^2 = 20,$$

and $\left(\sqrt{145}\right)^2 = 145$.

Thus

$a^2 + b^2 = c^2$ is equivalent to,

$$125 + 20 = 145,$$

$$145 = 145.$$

Therefore, triangle ABC is indeed a right triangle.

 Determine the constant A so that the lines $3x - 4y = 12$ and $Ax + 6y = -9$ are parallel.

 If two non-vertical lines are parallel, their slopes are equal. Thus the lines $Ax + By + C = 0$ and $Ax + By + D = 0$ are parallel (since both have slope $= -\dfrac{A}{B}$). We are given two lines:

$$3x - 4y = 12 \tag{1}$$

$$Ax + 6y = -9 \tag{2}$$

We must make the coefficients of y the same for both equations in order to equate the coefficients of x. Multiply (1) by $-\dfrac{3}{2}$ to obtain:

$$-\frac{3}{2}(3x - 4y) = -\frac{3}{2}(12) \tag{3}$$

$$-\frac{9}{2}x + 6y = -18$$

$$Ax + 6y = -9$$

Transpose the constant terms of (3) and (2) to the other side.

Adding 18 to both sides, $\dfrac{-9}{2}x + 6y + 18 = 0$ $\tag{4}$

Adding 9 to both sides, $Ax + 6y + 9 = 0$ $\tag{5}$

(4) and (5) will now be parallel if the coefficients of the x-terms are the same. Thus the constant A is $-\dfrac{9}{2}$. Then equation (5) becomes $-\dfrac{9}{2}x + 6y + 9 = 0$. We can also express (5) in its given form,

$$Ax + 6y = -9 \quad \text{or} \quad -\frac{9}{2}x + 6y = -9.$$

We also can write it in a form that has the same coefficient of x as (1), which clearly shows that they have equal slopes.

$$3x - 4y = 12$$

$$-\frac{9}{2}x + 6y = -9$$

Multiply the second equation by $-\frac{2}{3}$ to obtain a coefficient of x equal to 3.

$$-\frac{2}{3}\left(-\frac{9}{2}x + 6y\right) = -\frac{2}{3}(-9)$$

$$3x - 4y = 6$$

Now equations (1), $3x - 4y = 12$ and the equation $3x - 4y = 6$ are parallel since the coefficients of x and y are identical.

2.4 Symmetry

Two points P_1 and P_2 are said to be symmetric with respect to line l, if l is the perpendicular bisector of $\overline{P_1P_2}$ (See Figure 2.9.)

A figure is symmetric with respect to line l if for every point P_1 on the figure there is another point P_2 on the figure such that P_1 and P_2 are symmetric relative to l. Intuitively, this means that one portion of the figure is the mirror image of the other portion of the figure. (See Figure 2.10.)

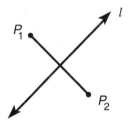

Figure 2.9

For the figure below, l is a symmetry line.

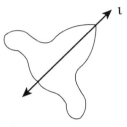

Figure 2.10

Two points P_1 and P_2 are symmetric with respect to Q when Q is the midpoint of $\overline{P_1 P_2}$. A figure is said to be symmetric relative to a point Q if for every point P_1 on the figure, there is a point P_2 on the figure such that P_1 and P_2 are symmetric relative to Q.

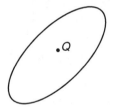

Figure 2.11

For the figure above, Q is a symmetry point.

When graphing equations, it is often valuable to know about symmetry points and lines. It is particularly important to know whether a given equation will have a graph which is symmetric to the origin, the x-axis, the y-axis, or the line $y = x$. A table follows which describes techniques for finding lines and points of symmetry.

Replace...	...to check for symmetry w/r to:
(x, y) by $(-x, -y)$	origin
(x, y) by $(x, -y)$	x – axis
(x, y) by $(-x, y)$	y – axis
(x, y) by (y, x)	line $y = x$

Table 2.1

Consider the equation $y = x^2$. Since $(-x)^2 = x^2$, the y-axis is a line of symmetry for the graph of this equation. The graph of this equation follows.

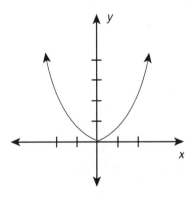

Figure 2.12

Next, consider the equation $y = x + x^3$. Then, $-y = -x + (-x)^3$ is equivalent to $y = x + x^3$, and the origin is a point of symmetry for the corresponding graph. The graph follows (Figure 2.13).

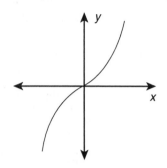

Figure 2.13

2.5   Intercepts and Asymptotes

The x-intercept and y-intercept for a line were described in Section 2.3. These terms have a similar meaning for the graph of a curve. Specifically, the second coordinates of points where a curve crosses the y-axis are called the y-intercepts, and the first coordinates of points where the curve crosses the x-axis are called the x-intercepts.

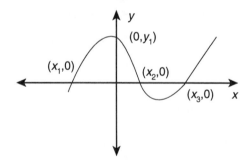

Figure 2.14

For the curve pictured above, the x-intercepts are x_1, x_2, and x_3 while the only y-intercept is y_1.

When the graph of an equation approaches a line, but never touches it, the line is said to be an asymptote. In most cases, these lines are vertical or horizontal and are called vertical asymptotes and horizontal asymptotes. In the picture below, the dotted lines are asymptotes.

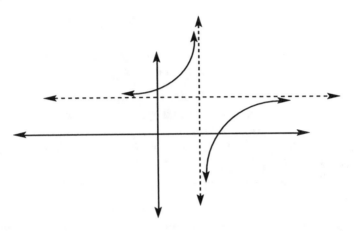

Figure 2.15

Problem Solving Example:

 Investigate each of the following equations, determine the intercepts and symmetry, and draw the graph.

(1) $x^2 + y^2 = 9$

(2) $2x^2y - x^2 - y = 0$

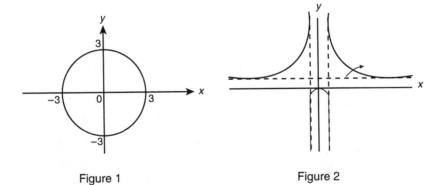

Figure 1   Figure 2

(1) $x^2 + y^2 = 9$

intercepts: $x = 0$ $y = \pm3$ (y-intercepts)

$y = 0$ $x = \pm3$ (x-intercepts)

symmetry: The locus is symmetric with respect to the coordinate axes and the origin.

The locus exists for all values of x where $x \geq -3$ and $x \leq 3$, and all values of y where $y \geq -3$ and $y \leq 3$. The locus consists of one closed piece. Various values of x were chosen and the values of y calculated as in the table shown below. The graph was then drawn, as shown in Figure 1.

x	-3	-2	-1	0	1	2	3
y	0	$\pm\sqrt{5}$	$\pm\sqrt{8}$	±3	$\pm\sqrt{8}$	$\pm\sqrt{5}$	0

(2) $2x^2y - x^2 - y = 0$

intercepts: $x = 0$ $y = 0$ (y–intercept)

$y = 0$ $x = 0$ (x-intercept)

symmetry: The locus is symmetric with respect to the y-axis.

extent: $y = \dfrac{x^2}{2x^2 - 1}$, $x = \pm\sqrt{\dfrac{y}{2y - 1}}$

The locus exists for all $x \neq \pm\sqrt{\dfrac{1}{2}}$ (this comes from the inequality $2x^2 - 1 \neq 0$), and for all y such that $\dfrac{y}{2y - 1} \geq 0$. Find the value of y for which the numerator is equal to zero, that is $y = 0$, and the value of y for which the denominator is zero, $y = \dfrac{1}{2}$. The locus exists for $y \leq 0$ and for $y > \dfrac{1}{2}$, since in these intervals the numerator and denominator have the same sign.

Asymptotes: $x = \pm\sqrt{\dfrac{1}{2}}, \quad y = \dfrac{1}{2}$

After choosing various values for x, and solving for y we form a table as follows:

x	0	±.5	±.7	±.8	±1	±5	±10
y	0	−.5	−24.5	2.29	1	.51	.502

The graph is seen in Figure 2.

2.6 Parametric Equations

The usual method for describing a curve is to give a single equation relating the variables x and y. However, sometimes it is more convenient to use two equations that express x and y in terms of a third variable. In such a case, the third variable is called a parameter, and the equations are called parametric equations. For example,

$$x = \cos\theta \quad y = \sin\theta$$

is an example of parametric equations with θ as the parameter. (We assume in this section and the next some understanding of trigonometry. See Chapter 6 for essentials of trigonometry.) It is quite easy to show that the corresponding graph is a circle, and, more specifically, the equations are equivalent to the equation:

$$x^2 + y^2 = 1.$$

A cycloid is defined to be the curve traced by a fixed point on a circle when the circle rolls on a line. With appropriate choice of the coordinate system, here is the graph resulting from rolling a circle of radius a. (See Figure 2.16.)

The parametric equations for the cycloid are:

$$x = a(\theta - \sin\theta) \quad \text{and} \quad y = a(1 - \cos\theta),$$

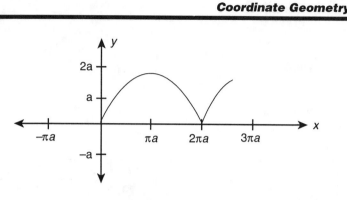

Figure 2.16

where θ must be measured in radians. It is possible to eliminate the parameter θ by a substitution method, but the resulting equation is quite complicated and so for this reason, the parametric equations are used almost invariably for a cycloid.

Problem Solving Examples:

 Graph the parametric equations

(1) $\begin{cases} x = 2t \\ y = t + 4 \end{cases}$

(2) $\begin{cases} x = \cos\theta \\ y = \sin\theta \end{cases}$

(3) $\begin{cases} x = 4t \\ y = \dfrac{2}{t} \end{cases}$

A (1) One of the methods for graphing these kinds of equations is to change them to regular equations without parameter. For instance, t in

(1) $\begin{cases} x = 2t & \qquad (1) \\ y = t + 4 & \qquad (2) \end{cases}$

is eliminated as the following:

from equation (1) $t = \dfrac{x}{2}$.   Substituting in equation (2),

$y = \dfrac{x}{2} + 4$, $x - 2y + 8 = 0$,   which is a straight line

whose graph is easily constructed by choosing two points on the line, as shown in Figure 1.

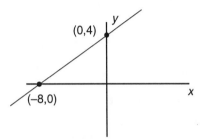

Figure 1

Note that another method for graphing would be using the table as shown below.

x	-8	-6	-4	-2	0	2
y	0	1	2	3	4	5

(2) $\begin{cases} x = \cos\theta & (3) \\ y = \sin\theta & (4) \end{cases}$

Squaring equations (3) and (4) and adding them together, one obtains

$$\begin{cases} x^2 = \cos^2\theta \\ y^2 = \sin^2\theta \end{cases}, \qquad x^2 + y^2 = \cos^2\theta + \sin^2\theta = 1$$

Hence, the graph is a circle of radius 1, with center $O(0,0)$.

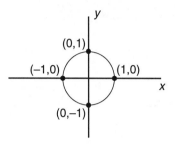

Figure 2

(3) $\begin{cases} x = 4t & (5) \\ y = \dfrac{2}{t} & (6) \end{cases}$

From equation (6), $t = \dfrac{2}{y}$. Substituting in equation (5),

$x = \dfrac{8}{y}$, $xy - 8 = 0$.

The graph is shown in Figure 3.

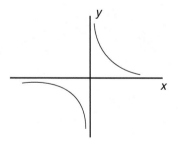

Figure 3

 Show that $x = 5\cos\theta$ and $y = 3\sin\theta$ satisfies the equation

$\dfrac{x^2}{25} + \dfrac{y^2}{9} = 1.$

A If $x = 5\cos\theta$, $y = 3\sin\theta$ satisfies the question

$$\frac{x^2}{25} + \frac{y^2}{9} = 1,$$

then we will obtain an identity when we substitute these values of x and y into the equation. Doing this, we find:

$$\frac{(5\cos\theta)^2}{25} + \frac{(3\sin\theta)^2}{9} = \frac{25\cos^2\theta}{25} + \frac{9\sin^2\theta}{9}$$
$$\cos^2\theta + \sin^2\theta = 1$$
$$1 = 1$$

Since this is an identity, the given values of x and y satisfy the given equation.

2.7 Polar Coordinates

In the rectangular coordinate system, a point is located by its distances from the x-axis and the y-axis. In the polar coordinate system, a fixed point O is selected as the origin and a fixed half line $\overrightarrow{OA}$ is chosen as the polar axis. Then the polar coordinates of a point P are (r, θ), where θ is the measure of the angle that $\overline{OP}$ makes with $\overline{OA}$, and r is the distance from O to P.

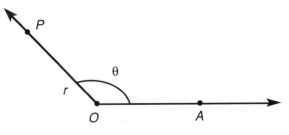

Figure 2.17

Then, when graphing the point $P(r, \theta)$

(1) using the polar axis as the initial side, lay off an angle of measure θ (counterclockwise if θ is positive and clockwise if θ is negative), and

(2) on the terminal side of this angle, measure off a segment of length r, if r is positive, and if r is negative, extend the terminal side of the angle past O and measure a segment of length $|r|$ along the extended side.

In practice, graphing in the polar coordinate system is usually done on special polar coordinate graph paper. An example follows.

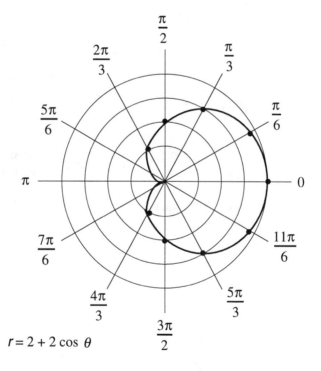

$r = 2 + 2 \cos \theta$

Figure 2.18

The heart-shaped figure is the graph of $r = 2 + 2 \cos \theta$ and is called a cardioid.

Problem Solving Examples:

 Draw the graph of $r = 2 \cos \theta$.

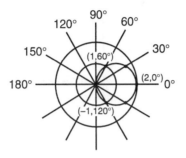

We assign values to θ and find the corresponding values of r, giving the following table:

θ	$\cos\theta$	$r = 2\cos\theta$
$0°$	1	2
$30°$	.87	1.74
$60°$	.5	1
$90°$	0	0
$120°$	−.5	−1
$150°$	−.87	−1.74
$180°$	−1	−2

Values from 180° to 360° give the same points.

We then plot the points (r, θ) and draw a smooth curve through them. We get the graph of the figure. The equation which defines the path of P may involve only one of the variables (r, θ). In that case the variable which is not mentioned may have any and all values.

Graph $r = \sin 3\theta$.

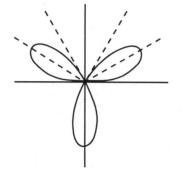

 We graph the figure in five steps. First, we obtain the equation for r in term θ.

(1) $r = \sin 3\theta$

Second, we find the boundary values of r and the symmetries. Note r will be at its greatest value, or most negative value, for

$3\theta = \frac{\pi}{2}, \frac{3\pi}{2}, \frac{5\pi}{2}$, etc. Corresponding values of θ are,

$\frac{\pi}{6}, \frac{\pi}{2}$ (negative r), $\frac{5\pi}{6}$, etc. (Subsequent values repeat earlier ones.)
Rays from the origin of those angles are lines of symmetry for the given equation (see figure).

Third, we find the intercepts. The pole intercepts are found by

setting $r = 0$; (0,0), $(0, \pm \frac{\pi}{3})$, and $(0, \pm \frac{2\pi}{3})$. The polar axis intercepts are found by setting $\theta = 0$, and π: (0,0) and (0, π). The 90°–

axis intercepts are found by setting $\theta = \pm \frac{\pi}{2}$: $(-1, \frac{\pi}{2})$ and $(1, -\frac{\pi}{2})$.
[Note that these points are the same.]

Fourth, we find out the general behavior of the curve. Recall the shape of $\sin \theta$ in rectangular coordinates.

θ	3θ	$r = \sin 3\theta$
Increasing from	Increasing from	Changing from
$-\dfrac{\pi}{2}$ to $-\dfrac{\pi}{3}$	$-\dfrac{3\pi}{2}$ to $-\pi$	1 to 0
$-\dfrac{\pi}{3}$ to $-\dfrac{\pi}{6}$	$-\pi$ to $-\dfrac{\pi}{2}$	0 to -1
$-\dfrac{\pi}{6}$ to 0	$-\dfrac{\pi}{2}$ to 0	-1 to 0
0 to $\dfrac{\pi}{6}$	0 to $\dfrac{\pi}{2}$	0 to 1
$\dfrac{\pi}{6}$ to $\dfrac{\pi}{3}$	$\dfrac{\pi}{2}$ to π	1 to 0
$\dfrac{\pi}{3}$ to $\dfrac{\pi}{2}$	π to $\dfrac{3\pi}{2}$	0 to -1

As a final step, we solve for particular points to make the sketch as precise as possible.

θ	3θ	$r = \sin 3\theta$
$\dfrac{\pi}{18}$	$\dfrac{\pi}{6}$	$\dfrac{1}{2} = .500$
$\dfrac{\pi}{12}$	$\dfrac{\pi}{4}$	$\dfrac{\sqrt{2}}{2} = .707$
$\dfrac{\pi}{9}$	$\dfrac{\pi}{3}$	$\dfrac{\sqrt{3}}{2} = .866$
$\dfrac{\pi}{6}$	$\dfrac{\pi}{2}$	$1 = 1.00$
$\dfrac{2\pi}{9}$	$\dfrac{2\pi}{3}$	$\dfrac{\sqrt{3}}{2} = +.866$
$\dfrac{\pi}{4}$	$\dfrac{3\pi}{4}$	$\dfrac{\sqrt{2}}{2} = .707$
$\dfrac{5\pi}{18}$	$\dfrac{5\pi}{6}$	$\dfrac{1}{2} = .500$
$\dfrac{\pi}{3}$	π	0

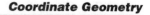
This curve is an example of a rose petal curve. The general equations of such a curve are

$$r = a \sin(n\theta) \qquad \text{and} \quad r = a \cos(n\theta)$$

where n is a positive integer.

The number of leaves of the curve is equal to n if n is an odd integer. If n is even, the number of leaves is $2n$. (See figure.)

 Transform the equation $r = 2\cos\theta$ to rectangular coordinates.

$$r = \sqrt{x^2 + y^2}, \quad \cos\theta = \frac{x}{r} = \frac{x}{\sqrt{x^2 + y^2}}$$

$$r = 2\cos\theta$$

$$\sqrt{x^2 + y^2} = \frac{2x}{\sqrt{x^2 + y^2}}$$

$$x^2 + y^2 = 2x$$

Quiz: Coordinate Geometry

1. In the given figure, the area of the triangle ABC is

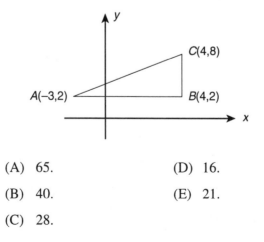

(A) 65.   (D) 16.

(B) 40.   (E) 21.

(C) 28.

2. Which of the following is the equation of the line that passes through the points (1,–4) and (–3,4)?

   (A) $y = -\frac{1}{2}x - 1$.

   (B) $y = -2 - 7$.

   (C) $y = 2x - 6$.

   (D) $y = \frac{1}{2}x + 3$.

   (E) $y = -2x - 2$.

3. Which of the following equations represents the line that is perpendicular to $2x + 3y = 4$ and passes through the point (–1,3)?

   (A) $3x - 2y = -9$.

   (B) $-3x + 2y = -9$.

   (C) $2x - y = -1$.

   (D) $2x - 3y = -11$.

   (E) $x - 3y = -4$.

4. Which of the following is symmetric with respect to the origin, the x-axis, the y-axis, and the line $y = x$?

   (A) $x^2 + \frac{y^2}{3} = 4$.

   (B) $x^2 + y^2 = 14$.

   (C) $\frac{x^2}{2} + \frac{y^2}{3} = 5$.

   (D) $x^3 + y^3 = -3$.

   (E) $\frac{x^4}{2} + \frac{y^2}{3} = 5$.

5. Which of the following is NOT symmetric with respect to either the origin, the x-axis or the y-axis?

   (A) $y = 3x^3 + x$.

   (B) $x = y^4 + y^2$.

   (C) $y = x^2 + 2x + 1$.

   (D) $y = -x + 1$.

   (E) $y = 4x^6$.

6. Given the equation $\dfrac{x^2}{4} + \dfrac{y^2}{6} = 1$ find all x and y intercepts.

    (A) x-intercepts: ±2, y-intercepts: $\pm\sqrt{6}$.

    (B) x-intercepts: 2, y-intercepts: $\sqrt{6}$.

    (C) x-intercepts: ±2, y-intercepts: ±3.

    (D) x-intercepts: ±2, y-intercepts: $\pm 2\sqrt{3}$.

    (E) x-intercepts: $\pm\sqrt{6}$, y-intercepts: ±2.

7. The vertical asymptote(s) of $y = \dfrac{2x}{x^2 - 2x - 3}$ are:

    (A) $y = 0$.   (D) $y = -1$ and $y = 3$.

    (B) $x = -1$ and $x = 3$.   (E) $x = 1$ and $x = -3$.

    (C) $x = 0$.

8. Which of the following equations is the rectangular representation for the parametric equations $\begin{cases} x = t + 3 \\ y = t^2 + 1 \end{cases}$?

    (A) $y = x^2 + 4$.   (D) $y = x^2 + 6x + 10$.

    (B) $y = x^2 - 6x + 10$.   (E) $y = x^2 + x + 3$.

    (C) $y = x^2 + x + 4$.

9. One endpoint of a line segment is (5,–3). The midpoint is (–1,6). What is the other endpoint?

    (A) (7,3).   (D) (–2,1.5).

    (B) (2,1.5).   (E) (–7,12).

    (C) (–7,15).

10. Where does the point $(-3, -\frac{\pi}{3})$ lie when plotted in a rectangular coordinate system?

    (A) on the x-axis.

    (B) Quandrant I.

    (C) Quandrant II.

    (D) Quandrant III.

    (E) Quandrant IV.

ANSWER KEY

1.	(E)	6.	(A)
2.	(E)	7.	(B)
3.	(A)	8.	(B)
4.	(B)	9.	(C)
5.	(C)	10.	(C)

CHAPTER 3

Fundamental Algebraic Topics

3.1 The Arithmetic of Polynomials

A monomial is a real number, or the product of a real number with one or more variables having whole number exponents. A polynomial is a finite sum of monomials. The sum or difference of two polynomials is found by combining like terms. The product of two polynomials can be determined by using the distributive property. However, here are some products which should be remembered.

1. $(x + y)(x - y) = x^2 - y^2$

2. $(x + y)^2 = x^2 + 2xy + y^2$

3. $(x - y)^2 = x^2 - 2xy + y^2$

4. $(x + y)^3 = x^3 + 3x^2y + 3xy^2 + y^3$

5. $(x - y)^3 = x^3 - 3x^2y + 3xy^2 - y^3$

6. $(x - y)(x^2 + xy + y^2) = x^3 - y^3$

7. $(x + y)(x^2 - xy + y^2) = x^3 + y^3$

8. $(ax + b)(cx + d) = acx^2 + (ad + bc)x + bd$

The process for finding the quotient of two polynomials is somewhat more complicated. This process is illustrated with the example below.

EXAMPLE

Find:

$$\frac{2x^2 - 7x + 8}{x - 2}$$

$$\begin{array}{r} 2x - 3 \\ x - 2 \overline{\smash{\big)}\, 2x^2 - 7x + 8} \\ \underline{2x^2 - 4x} \\ -3x + 8 \\ \underline{-3x + 6} \\ 2 \end{array}$$

Thus,

$$\frac{2x^2 - 7x + 8}{x - 2} = 2x - 3 + \frac{2}{x - 2}$$

Problem Solving Examples:

 From the sum of $6x^2 + 4xy - 8y^2 - 11$ and $3x^2 - 4y^2 + 8 + 5xy$ subtract $xy - 10 - 5x^2 + 7y^2$.

 First find the sum of $6x^2 + 4xy - 8y^2 - 11$ and $3x^2 - 4y^2 + 8 + 5xy$. Adding these two polynomials together:

$$\left(6x^2 + 4xy - 8y^2 - 11\right) + \left(3x^2 - 4y^2 + 8 + 5xy\right)$$

$$= 6x^2 + 4xy - 8y^2 - 11 + 3x^2 - 4y^2 + 8 + 5xy$$

Grouping like terms together,

$$= \left(6x^2 + 3x^2\right) + \left(4xy + 5xy\right) + \left(-8y^2 - 4y^2\right) + \left(-11 + 8\right)$$
$$= 9x^2 + 9xy + \left(-12y^2\right) + \left(-3\right)$$
$$= 9x^2 + 9xy - 12y^2 - 3 \tag{1}$$

Now subtract $xy - 10 - 5x^2 + 7y^2$ from the resultant sum, which is (1), or $9x^2 + 9xy - 12y^2 - 3$. Then,

$$\left(9x^2 + 9xy - 12y^2 - 3\right) - \left(xy - 10 - 5x^2 + 7y^2\right) =$$
$$= 9x^2 + 9xy - 12y^2 - 3 - xy + 10 + 5x^2 - 7y^2 \tag{2}$$

Grouping like terms together, equation (2) becomes:

$$\left(9x^2 + 9xy - 12y^2 - 3\right) - \left(xy - 10 - 5x^2 + 7y^2\right)$$
$$= \left(9x^2 + 5x^2\right) + \left(9xy - xy\right) + \left(-12y^2 - 7y^2\right) + \left(-3 + 10\right)$$
$$= 14x^2 + 8xy + \left(-19y^2\right) + 7$$
$$= 14x^2 + 8xy - 19y^2 + 7,$$

which is the final answer.

Subtract $3x^4y^3 + 5x^2y - 4xy + 5x - 3$ from the polynomial $5x^4y^3 - 3x^2y + 7$.

A

$$\left(5x^4y^3 - 3x^2y + 7\right) - \left(3x^4y^3 + 5x^2y - 4xy + 5x - 3\right)$$
$$= \left(5x^4y^3 - 3x^2y + 7\right) + \left(-3x^4y^3 - 5x^2y + 4xy - 5x + 3\right)$$
$$= 5x^4y^3 + \left(-3x^4y^3\right) - 3x^2y + \left(-5x^2y\right) + 4xy - 5x + 7 + 3$$
$$= 2x^4y^3 - 8x^2y + 4xy - 5x + 10$$

The column form may also be used for subtraction. Here we align the like terms and subtract the coefficients.

$$5x^4y^3 - 3x^2y \qquad\qquad + 7$$
$$\underline{-\left[3x^4y^3 + 5x^2y - 4xy + 5x - 3\right]}$$
$$2x^4y^3 - 8x^2y + 4xy - 5x + 10$$

 Simplify $(5ax + by)(2ax - 3by)$.

 The following formula can be used to simplify the given expression:

$$\left(N_1 + N_2\right)N_3 = N_1N_3 + N_2N_3$$

where N_1, N_2 and N_3 are any three numbers. Note that the distributive property is used in this formula. N_1 is replaced by $5ax$ and by replaces N_2. Also, N_3 is replaced by $(2ax - 3by)$. Therefore:

$$(5ax + by)(2ax - 3by) = 5ax(2ax - 3by) + by(2ax - 3by)$$

Use the distributive property to simplify the right side of the above equation:

$$(5ax + by)(2ax - 3by) = 5ax(2ax) + 5ax(-3by) + by(2ax) + by(-3by)$$
$$= 10a^2x^2 - 15abxy + 2abxy - 3b^2y^2$$
$$= 10a^2x^2 - 13abxy - 3b^2y^2$$

It is often convenient to arrange the two factors vertically as we do in ordinary arithmetic. Hence, the problem is an ordinary multiplication problem.

$$5ax + by$$
$$\underline{\times \quad 2ax - 3by}$$
$$-15abxy - 3b^2y^2$$
$$\underline{+10a^2x^2 + 2abxy}$$
$$10a^2x^2 - 13abxy - 3b^2y^2$$

Note that this answer (i.e., product) is the same as the answer obtained above.

 Divide $3x^5 - 8x^4 - 5x^3 + 26x^2 - 33x + 26$ by $x^3 - 2x^2 - 4x + 8$

A To divide a polynomial by another polynomial we set up the divisor and the dividend as shown below. Then we divide the first term of the divisor into the first term of the dividend. We multiply the quotient from this division by each term of the divisor, and subtract the products of each term from the dividend. We then obtain a new dividend. Use this dividend, and again divide by the first term of the divisor, and repeat all steps again until we obtain a remainder which is of degree lower than that of the divisor or = zero. Following this procedure we obtain:

$$
\begin{array}{r}
3x^2 - 2x + 3 \\
x^3 - 2x^2 - 4x + 8 \overline{\smash{\big)}\ 3x^5 - 8x^4 - 5x^3 + 26x^2 - 33x + 26} \\
\underline{3x^5 - 6x^4 - 12x^3 + 24x^2} \\
-2x^4 + 7x^3 + 2x^2 - 33x + 26 \\
\underline{-2x^4 + 4x^3 + 8x^2 - 16x} \\
3x^3 - 6x^2 - 17x + 26 \\
\underline{3x^3 - 6x^2 - 12x + 24} \\
-5x + 2
\end{array}
$$

Thus, the quotient is $3x^2 - 2x + 3$ and the remainder is $-5x + 2$.

Multiply: $(x+2y)(x-2y)(x^2+4y^2)$.

Here we use the factoring formula $a^2 - b^2 = (a-b)(a+b)$ to rewrite the product $(x+2y)(x-2y)$:

$(x+2y)(x-2y) = (x)^2 - (2y)^2$ difference of two squares.

$$= x^2 - 4y^2.$$

Hence, $(x+2y)(x-2y)(x^2+4y^2) = (x^2-4y^2)(x^2+4y^2)$

(1)

Now, again use the factoring formula given above to rewrite the right side of equation (1) in which $x^2 = a$ and $4y^2 = b$. Hence,

$$(x^2-4y^2)(x^2+4y^2) = (x^2)^2 - (4y^2)^2$$

$$= x^4 - 4^2 y^4 \quad \text{since} \quad (a^x)^y = a^{x \cdot y}$$

$$= x^4 - 16y^4$$

Hence, equation (1) becomes:

$(x+2y)(x-2y)(x^2+4y^2) = x^4 - 16y^4$.

3.2 Factoring

Factoring is the inverse operation of finding products. In this situation, a polynomial is given and the student is to express the polynomial as the product of at least two polynomials. In the process, it is imperative that the student memorize the product types given in the previous section. Here are some procedures which make the process easier.

(1) Look for and factor out any common monomial factors.

(2) Look for any of the product types and apply any identified patterns.

(3) Try a trial-and-error approach.

(4) Try grouping terms differently.

Here are some examples which illustrate these procedures.

EXAMPLE

Factor

$$x^2 - 16$$
$$x^2 - 16 = (x - 4)(x + 4)$$

Factor

$$4x^2 + 12xy + 9y^2$$
$$4x^2 + 12xy + 9y^2 = (2x + 3y)^2$$

Factor

$$2x^3 - 16y^3$$
$$2x^3 - 16y^3 = 2(x^3 - 8y^3)$$
$$= 2(x - 2y)(x^2 + 2xy + 4y^2)$$

Factor

$$2x^2 - xy - 6y^2$$
$$2x^2 - xy - 6y^2 = (2x + 3y)(x - 2y)$$

Factor

$$ax + ay + 3x + 3y$$
$$ax + ay + 3x + 3y = a(x + y) + 3(x + y)$$
$$= (a + 3)(x + y)$$

Problem Solving Examples:

Factor (a) $4a^2b - 2ab$

(b) $9ab^2c^3 - 6a^2c + 12ac$

(c) $ac + bc + ad + bd$

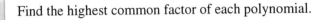 Find the highest common factor of each polynomial.

(a)
$$4a^2b = 2 \times 2 \times a \times a \times b$$
$$2ab = 2 \times a \times b$$

The highest common factor of the two terms is therefore $2ab$. Hence,

$$4a^2b - 2ab = 2ab(2a - 1).$$

(b)
$$9ab^2c^3 = 3 \times 3 \times a \times b \times b \times c \times c \times c$$
$$6a^2c = 3 \times 2 \times a \times a \times c$$
$$12ac = 3 \times 2 \times 2 \times a \times c$$

The highest common factor of the three terms is $3ac$. Then,

$$9ab^2c^3 - 6a^2c + 12ac = 3ac(3b^2c^2 - 2a + 4)$$

(c) An expression may sometimes be factored by grouping terms having a common factor and thus getting new terms containing a common factor. The type form for this case is $ac + bc + ad + bd$, because the terms ac and bc have the common factor c, and ad and bd have the common factor d. Then,

$$ac + bc + ad + bd = c(a + b) + d(a + b)$$

Factoring out $(a + b)$, we obtain:

$$= (a + b)(c + d).$$

 Factor $xy - 3y + y^2 - 3x$ completely.

 Note that the first and last terms have a common factor of x. Also note that the second and third terms have a common factor of y. Hence, group the x and y terms together and factor out the x and y from their respective two terms. Therefore,

$$xy - 3y + y^2 - 3x = (xy - 3x) + (-3y + y^2)$$

Since $(-3y + y^2) = (y^2 - 3y),$

$$xy - 3y + y^2 - 3x = (xy - 3x) + (y^2 - 3y)$$
$$= x(y - 3) + y(y - 3)$$

Now factor out the common factor $(y - 3)$ from both terms:

$$xy - 3y + y^2 - 3x = (x + y)(y - 3).$$

 Factor $125m^3n^6 - 8a^3$.

A $125 = 5 \times 5 \times 5 = 5^3$. Also since $a^{xy} = (a^x)^y$, $n^6 = n^{2 \times 3} = (n^2)^3$.

Thus $125m^3n^6 = 5^3m^3(n^2)^3$. Since $a^x b^x c^x = (abc)^x$,

$5^3m^3(n^2)^3 = (5mn^2)^3$.

$8 = 2 \times 2 \times 2 = 2^3$,

thus $8a^3 = 2^3a^3 = (2a)^3$.

Now $125m^3n^6 - 8a^3 = (5mn^2)^3 - (2a)^3$,

which is the difference of two cubes.

Apply the formula for the difference of

two cubes $x^3 - y^3 = (x - y)(x^2 + xy + y^2)$,

replacing x by $5mn^2$ and y by $2a$.

Hence,

$125m^3n^6 - 8a = (5mn^2 - 2a)[(5mn^2)^2 + 5mn^2(2a) + (2a)^2].$

$= (5mn^2 - 2a)(25m^2n^4 + 10amn^2 + 4a^2)$

Factor $(x + y)^3 + z^3$.

The given expression is the sum of two cubes. The formula for the sum of two cubes can be used to factor the given expression.

This formula is:

$$a^3 + b^3 = (a + b)(a^2 - ab + b^2).$$

Using this formula and replacing a by $x + y$ and b by z:

$$(x + y)^3 + z^3 = [(x + y) + z][(x + y)^2 - (x + y)z + z^2]$$

Factor (a) $a^4 + 4a^2 + 4$

(b) $9a^2 - 6ab^2 + b^4.$

The first example is a trinomial which is a perfect square, in the form:

$$x^2 + 2xy + y^2 = x^2 + xy + xy + y^2 = (x + y)(x + y) = (x + y)^2.$$

For example (a), replace x by a^2 and y by 2 to obtain

$$a^4 + 4a^2 + 4 = (a^2)^2 + 2 \times a^2 \times 2 + 2^2 = (a^2 + 2)^2,$$

The second example is a trinomial perfect square whose form is:

$$x^2 - 2xy + y^2 = x^2 - xy - xy + y^2 = (x - y)(x - y) = (x - y)^2$$

For example (b) replace x by 3a and y by b^2 to obtain:

$$9a^2 - 6ab^2 + b^4 = (3a)^2 - 2(3a)\left(b^2\right) + \left(b^2\right)^2$$
$$= \left(3a - b^2\right)\left(3a - b^2\right)$$
$$= \left(3a - b^2\right)^2$$

3.3   Rational Expressions

A rational expression is an expression of the type $\dfrac{P(x)}{Q(x)}$, where

$P(x)$ and $Q(x)$ are both polynomials in the variable x and $Q(x) \neq 0$.

Also, $\dfrac{P(x)}{Q(x)}$ is in simplest form whenever the only common factors

of $P(x)$ and $Q(x)$ are 1 and -1.

Here are some important generalizations concerning rational expressions.

$$\frac{P(x)}{Q(x)} = \frac{R(x)}{S(x)}$$

if and only if $P(x) \times S(x) = Q(x) \times R(x)$

$$\frac{P(x)}{Q(x)} = \frac{P(x) \times S(x)}{Q(x) \times S(x)}$$

$$\frac{P(x)}{Q(x)} + \frac{R(x)}{Q(x)} = \frac{P(x) + R(x)}{Q(x)}$$

$$\frac{P(x)}{Q(x)} - \frac{R(x)}{Q(x)} = \frac{P(x) - R(x)}{Q(x)}$$

$$\frac{P(x)}{Q(x)} \times \frac{R(x)}{S(x)} = \frac{P(x) \times R(x)}{Q(x) \times S(x)}$$

$$\frac{P(x)}{Q(x)} \div \frac{R(x)}{S(x)} = \frac{P(x)}{Q(x)} \times \frac{S(x)}{R(x)}$$

Here are some examples which illustrate the application of these generalizations.

EXAMPLE

Simplify:

$$\frac{x^2 + 3x - 10}{x^2 - 4} \div \frac{x^2 + 2x - 15}{x^2 - 2x - 3}$$

$$\frac{x^2 + 3x - 10}{x^2 - 4} \div \frac{x^2 + 2x - 15}{x^2 - 2x - 3} = \frac{x^2 + 3x - 10}{x^2 - 4} \times \frac{x^2 - 2x - 3}{x^2 + 2x - 15}$$

$$= \frac{(x+5)(x-2)(x-3)(x+1)}{(x-2)(x+2)(x+5)(x-3)}$$

$$= \frac{x+1}{x+2}$$

Simplify:

$$\frac{x-2}{x^2-1} - \frac{x+3}{x^2+3x+2} - \frac{-3}{x^2+x-2}$$

$$\frac{x-2}{x^2-1} - \frac{x+3}{x^2+3x+2} - \frac{-3}{x^2+x-2}$$

$$= \frac{x-2}{(x-1)(x+1)} - \frac{x+3}{(x+2)(x+1)} - \frac{-3}{(x+2)(x-1)}$$

$$= \frac{(x-2)(x+2)}{(x-1)(x+1)(x+2)} - \frac{(x+3)(x-1)}{(x+2)(x+1)(x-1)} - \frac{-3(x+1)}{(x+2)(x-1)(x+1)}$$

$$= \frac{(x^2-4)-(x^2+2x-3)-(-3x-3)}{(x-1)(x+1)(x+2)}$$

$$= \frac{x+2}{(x-1)(x+1)(x+2)}$$

$$= \frac{1}{(x-1)(x+1)}$$

Problem Solving Examples:

Combine into a single fraction in lowest terms.

(a) $$\frac{6(a+1)}{a+8} - \frac{3(a-4)}{a+8} - \frac{2(a+5)}{a+8}$$

(b) $$\frac{7x-3y+6}{x+y} - \frac{2(x-4y+3)}{x+y}$$

(c) $$\frac{5x+2}{x-6} - \frac{3(x+4)}{x-6} - \frac{x-7}{x-6}$$

Noting $\frac{a}{x} + \frac{b}{x} + \frac{c}{x} = \frac{a+b+c}{x}$ (where a, b, c are any real number and x any nonzero real number), we proceed to evaluate these expressions:

(a)

$$\frac{6(a+1)}{a+8} - \frac{3(a-4)}{a+8} - \frac{2(a+5)}{a+8} = \frac{6(a+1)-3(a-4)-2(a+5)}{a+8}$$

Distributing,

$$= \frac{6a+6-3a+12-2a-10}{a+8} = \frac{6a-3a-2a+6+12-10}{a+8}$$

$$= \frac{a+8}{a+8} = 1.$$

(b)

$$\frac{7x-3y+6}{x+y} - \frac{2(x-4y+3)}{x+y} = \frac{7x-3y+6-2(x-4y+3)}{x+y}$$

Distributing,

$$= \frac{7x-3y+6-2x+8y-6}{x+y} = \frac{7x-2x-3y+8y+6-6}{x+y}$$

$$= \frac{5x+5y}{x+y} = \frac{5(x+y)}{x+y} = 5.$$

(c)

$$\frac{5x+2}{x-6} - \frac{3(x+4)}{x-6} - \frac{x-7}{x-6} = \frac{5x+2-3(x+4)-(x-7)}{x-6}$$

Distributing,

$$= \frac{5x+2-3x-12-x+7}{x-6} = \frac{5x-3x-x+2-12+7}{x-6} = \frac{x-3}{x-6}$$

 Combine $\dfrac{3x+y}{x^2-y^2} - \dfrac{2y}{x(x-y)} - \dfrac{1}{x+y}$ into a single fraction.

A Fractions which have unlike denominators must be transformed into fractions with the same denominator before they may be combined. This identical denominator is the least common denominator (LCD), the least common multiple of the denominators of the fractions to be added. In the process of transforming the fractions to fractions with a common denominator, we make use of the fact that the numerator and denominator of a fraction may be multiplied by the same non-zero number without changing the value of the fraction. In our case the denominators are:

$$x^2 - y^2 = (x+y)(x-y), \, x(x-y), \text{ and } x+y.$$

Therefore the LCD is $x(x+y)(x-y)$, and we proceed as follows:

$$\frac{3x+y}{x^2-y^2} - \frac{2y}{x(x-y)} - \frac{1}{x+y} = \frac{3x+y}{(x+y)(x-y)} - \frac{2y}{x(x-y)} - \frac{1}{x+y}$$

$$= \frac{x(3x+y)}{x(x+y)(x-y)} - \frac{(x+y)2y}{(x+y)(x)(x-y)}$$

$$- \frac{x(x-y)}{x(x-y)(x+y)}$$

$$= \frac{3x^2+xy}{x(x+y)(x-y)} - \frac{2xy+2y^2}{x(x+y)(x-y)}$$

$$- \frac{x^2-xy}{x(x+y)(x-y)}$$

$$= \frac{3x^2+xy-\left(2xy+2y^2\right)-\left(x^2-xy\right)}{x(x+y)(x-y)}$$

$$= \frac{3x^2+xy-2xy-2y^2-x^2+xy}{x(x+y)(x-y)}$$

$$= \frac{2x^2-2y^2}{x(x+y)(x-y)} = \frac{2\left(x^2-y^2\right)}{x(x+y)(x-y)}$$

$$= \frac{2(x+y)(x-y)}{x(x+y)(x-y)} = \frac{2}{x}.$$

Reduce $\dfrac{4x-20}{50-2x^2}$ to the lowest terms.

 Factor the numerator and the denominator:

$$\frac{4x-20}{50-2x^2} = \frac{4(x-5)}{2(25-x^2)} = \frac{4(x-5)}{2(5-x)(5+x)}$$

Multiply the numerator and denominator by (-1) to reverse the sign of the factor (5-x) in the denominator. Then divide both the numerator and denominator by 2(x-5).

$$\frac{(-1)[4(x-5)]}{(-1)[2(5-x)(5+x)]} = \frac{-4(x-5)}{2(x-5)(5+x)}$$

Dividing, we obtain:

$$-\frac{2}{x+5}.$$

 Divide $\dfrac{y^2+y-20}{y-3}$ by $\dfrac{y^2-16}{y^2+y-12}$.

 Dividing by a nonzero polynomial is the same as multiplying by its reciprocal. That is,

$$\frac{y^2+y-20}{y-3} \div \frac{y^2-16}{y^2+y-12} = \frac{y^2+y-20}{y-3} \times \frac{y^2+y-12}{y^2-16}$$

Factor each numerator and denominator, where possible. Note

that $\quad y^2+y-20 = (y+5)(y-4)$

$$y^2+y-12 = (y+4)(y-3),$$

and $y^2-16 = y^2-4^2$, the difference of two squares. Using the formula for the difference of two squares, $(a^2-b^2)=(a-b)(a+b)$, replace a by y and b by 4 to obtain, $(y^2-16)=(y-4)(y+4)$.

Thus, $\dfrac{y^2+y-20}{y-3} \times \dfrac{y^2+y-12}{y-16}$

$$= \dfrac{(y+5)(y-4)}{y-3} \times \dfrac{(y+4)(y-3)}{(y-4)(y+4)} \qquad (1)$$

$$= \dfrac{(y+5)(y-4)(y+4)(y-3)}{(y-3)(y-4)(y+4)} \qquad (2)$$

$$= \dfrac{(y+5)(y-4)(y+4)(y-3)}{(y-4)(y+4)(y-3)} \qquad (3)$$

$$= y+5.$$

Note that in equation (2) we are dividing by $(y-3)(y-4)(y+4)$. If any of these factors equal 0, then we are dividing by zero, making our fraction invalid. Thus, in order to be certain we are proceeding correctly, we must establish the following restrictions:

$(y-3) \neq 0, \qquad (y-4) \neq 0, \qquad (y+4) \neq 0;$

thus, $y \neq 3, \quad y \neq 4, \quad y \neq -4,$

Therefore, $\dfrac{y^2+y-20}{y-3} \div \dfrac{y^2-16}{y^2+y-12} = y+5,$

and $y \neq 3, 4, -4.$

 Perform the indicated operation,

$$\dfrac{x^3-y^3}{x^2-5x+6} \times \dfrac{x^2-4}{x^2-2xy+y^2}$$

 We factor numerators and denominators to enable us to cancel terms.

$$x^3 - y^3$$

is the difference of two cubes. Thus we factor it applying the formula for the difference of two cubes,

$$a^3 - b^3 = (a-b)(a^2 + ab + b^2),$$

replacing a by x and b by y. Thus,

$$x^3 - y^3 = (x-y)(x^2 + xy + y^2).$$

$x^2 - 5x + 6$ is factored as $(x-2)(x-3)$.

$$x^2 - 4 = x^2 - 2^2,$$

the difference of two squares. Applying the formula for the difference of two squares,

$$a^2 - b^2 = (a+b)(a-b),$$

and replacing a by x and b by 2 we obtain,

$$x^2 - 4 = (x+2)(x-2).$$

$$x^2 - 2xy + y^2 = (x-y)(x-y).$$

Thus,

$$\frac{x^3 - y^3}{x^2 - 5x + 6} \times \frac{x^2 - 4}{x^2 - 2xy + y^2} = \frac{(x-y)(x^2 + xy + y^2)}{(x-2)(x-3)} \times \frac{(x+2)(x-2)}{(x-y)(x-y)}$$

$$= \frac{(x^2 + xy + y^2)(x-y)(x-2)(x+2)}{(x-3)(x-2)(x-y)(x-y)}$$

$$= \frac{(x^2 + xy + y^2)(x+2)}{(x-3)(x-y)}.$$

3.4 Radicals

The symbol "$\sqrt[n]{a}$" is used to represent *the* principal n^{th} root of a, and more specifically "$\sqrt{4}$" is used to represent the principal square root of 4. Since, for example,

$$2^2 = 4 \quad \text{and} \quad (-2)^2 = 4,$$

2 and -2 are candidates for $\sqrt{4}$. Thus, there is need for a definition which differentiates between these two options. A similar situation occurs for any even root. Also, since a real number raised to an even power cannot be negative, $\sqrt[n]{a}$ is not a real number when n is even and a is negative. When n is odd $\sqrt[n]{a}$ is meaningful for any real number a. Thus, there is need to differentiate between the even and odd cases for $\sqrt[n]{a}$. Here are the definitions for $\sqrt[n]{a}$.

For every real number a, $a \geq 0$, and for an even positive integer n,

$\sqrt[n]{a} = b$ if and only if $b^n = a$ and $b \geq 0$.

For every real number a, for every odd positive integer n,

$n > 1$, $\sqrt[n]{a} = b$ if and only if $b^n = a$.

Here are several important generalizations concerning radicals.

If a and b are real numbers, and if n is a positive integer such that $\sqrt[n]{a}$ and $\sqrt[n]{b}$ are real numbers, then

$$\sqrt[n]{ab} = \sqrt[n]{a}\sqrt[n]{b}$$

$$\sqrt[n]{\frac{a}{b}} = \frac{\sqrt[n]{a}}{\sqrt[n]{b}}$$

$$\sqrt[n]{a^n} = a \qquad \text{(when } n \text{ is odd)}$$

$$\sqrt[n]{a^n} = |a| \qquad \text{(when } n \text{ is even)}$$

These generalizations can be used to rewrite radical expressions in different ways. In this regard, a radical expression is said to be in standard (simplest) form if:

(1)   There are no polynomial factors in the radicand raised to a higher power than the index of the radical.

(2)   There are no fractions under the radical sign.

(3)   There are no radicals contained in the denominator.

(4)   The index of the radical is as small as possible.

Here are some examples where radicals are transformed to standard form.

EXAMPLE

(a) $\sqrt[5]{x^7} = \sqrt[5]{x^5} \sqrt[5]{x^2}$

$\qquad = x \sqrt[5]{x^2}$

(b) $\sqrt[4]{\dfrac{1}{x^3}} = \sqrt[4]{\dfrac{x}{x^4}}$

$\qquad = \dfrac{\sqrt[4]{x}}{\sqrt[4]{x^4}}$

$\qquad = \dfrac{\sqrt[4]{x}}{|x|}$

(c) $\dfrac{1}{4+\sqrt{3}} = \dfrac{1\left(4-\sqrt{3}\right)}{\left(4+\sqrt{3}\right)\left(4-\sqrt{3}\right)}$

$\qquad = \dfrac{4-\sqrt{3}}{16-3}$

$\qquad \dfrac{4-\sqrt{3}}{13}$

(d) $\sqrt[9]{x^3} = \left(x^3\right)^{\frac{1}{9}}$

$= x^{\frac{1}{3}}$

$= \sqrt[3]{x}$

Notice that in the last example, the generalization $\sqrt[n]{a} = a^{1/n}$ is used. This generalization is discussed in Section 7.2.

Find the indicated roots.

(a) $\sqrt[5]{32}$

(b) $\pm\sqrt[4]{625}$

(c) $\sqrt[3]{-125}$

(d) $\sqrt[4]{-16}$

A The following two laws of exponents can be used to solve these problems:

(1) $\left(\sqrt[n]{a}\right)^n = \left(a^{\frac{1}{n}}\right)^n = a^1 = a,$ and

(2) $\left(\sqrt[n]{a}\right)^n = \sqrt[n]{a^n}.$

(a) $\sqrt[5]{32} = \sqrt[5]{2^5} = \left(\sqrt[5]{2}\right)^5 = 2.$

This result is true because

$(2)^5 = 32,$ that is, $2 \times 2 \times 2 \times 2 \times 2 = 32.$

(b) $\sqrt[4]{625} = \sqrt[4]{5^4} = \left(\sqrt[4]{5}\right)^4 = 5.$ This result is true because $\left(5^4 = 625\right),$ that is, $5 \times 5 \times 5 \times 5 = 625.$

$-\sqrt[4]{625} = -\left(\sqrt[4]{5^4}\right) = -[5] = -5.$

(c) $\sqrt[3]{-125} = \sqrt[3]{(-5)^3} = (\sqrt[3]{-5})^3 = -5$. This result is true because $(-5)^3 = -125$, that is, $(-5)\times(-5)\times(-5) = -125$.

(d) There is no solution to $\sqrt[4]{-16}$ because any number raised to the fourth power is a positive number, that is, $N^4 = (N)\times(N)\times(N)\times(N)$ is a positive number $\neq$ a negative number, -16.

 Simplify $5\sqrt{12} + 3\sqrt{75}$

 Express 12 and 75 as the product of perfect squares if possible. Thus,

$12 = 4\times 3$ and $75 = 25\times 3$;

and $5\sqrt{12} + 3\sqrt{75} = 5\sqrt{4\times 3} + 3\sqrt{25\times 3}$.

Since

$$\sqrt{a\times b} = \sqrt{a}\times\sqrt{b}: \quad = \left[5\times\sqrt{4}\times\sqrt{3}\right] + \left[3\sqrt{25}\times\sqrt{3}\right]$$
$$= \left[(5\times 2)\sqrt{3}\right] + \left[(3\times 5)\sqrt{3}\right]$$
$$= 10\sqrt{3} + 15\sqrt{3}.$$

Using the distributive law:
$$= (10+15)\sqrt{3}$$
$$= 25\sqrt{3}.$$

 Simplify the quotient $\dfrac{\sqrt{x}}{\sqrt[4]{x}}$. Write the result in exponential notation.

Since $\sqrt[b]{n^a} = n^{\frac{a}{b}}$, the numerator and the denominator can be rewritten as:

$$\sqrt{x} = x^{\frac{1}{2}} \quad \text{and}$$

$$\sqrt[4]{x} = x^{\frac{1}{4}}$$

Therefore,

$$\frac{\sqrt{x}}{\sqrt[4]{x}} = \frac{x^{\frac{1}{2}}}{x^{\frac{1}{4}}} \tag{1}$$

According to the law of exponents which states that $\frac{n^a}{n^b} = n^{a-b}$, equation (1) becomes:

$$\frac{\sqrt{x}}{\sqrt[4]{x}} = \frac{x^{\frac{1}{2}}}{x^{\frac{1}{4}}}$$

$$= x^{\frac{1}{2}-\frac{1}{4}}$$

$$= x^{\frac{2}{4}-\frac{1}{4}}$$

$$= x^{\frac{1}{4}}$$

(We assume all variables are non-negative in the problems below.)

 Simplify: (a) $\sqrt{8x^3y}$

                      (b) $\sqrt{\dfrac{2a}{4b^2}}$

                      (c) $\sqrt[4]{25x^2}$.

 (a) $\sqrt{8x^3y}$ contains the perfect square $4x^2$. Factoring out $4x^2$ we obtain,

$$\sqrt{8x^3y} = \sqrt{4x^2 \times 2xy}.$$

Recall that $\sqrt{ab} = \sqrt{a} \times \sqrt{b}$. Thus,

$$= \sqrt{4x^2} \times \sqrt{2xy}$$
$$= \sqrt{4}\sqrt{x^2}\sqrt{2xy}$$
$$= 2x\sqrt{2xy}.$$

(b) $\sqrt{\dfrac{2a}{4b^2}}$ has a denominator that is a perfect square.

$$\sqrt{\frac{2a}{4b^2}} = \frac{\sqrt{2a}}{\sqrt{4b^2}}, \quad \text{since} \quad \sqrt{\frac{a}{b}} = \frac{\sqrt{a}}{\sqrt{b}}$$

$$\frac{\sqrt{2a}}{\sqrt{4}\sqrt{b^2}} = \frac{\sqrt{2a}}{2b}.$$

(c) $\sqrt[4]{25x^2}$ has a perfect square for the radicand.

$$\sqrt[4]{25x^2} = \sqrt[4]{(5x)^2}.$$

Recall that $\sqrt[4]{x} = \sqrt[2]{\sqrt[2]{x}}$; hence $\sqrt[4]{(5x)^2} = \sqrt[2]{\sqrt[2]{(5x)^2}}$. Now, since

$$\sqrt[2]{(5x)^2} = 5x, \quad = \sqrt[2]{5x}.$$

Radicals with the same index can be multiplied by finding the product of the radicands, the index of the product being the same as the factors.

 Rationalize $\dfrac{\sqrt[3]{3ax}}{\sqrt[3]{4a^2}}$.

 Multiply the numerator and the denominator by the radical $\sqrt[3]{(4a^2)^2}$ to eliminate the radical in the denominator:

$$\frac{\sqrt[3]{3ax}}{\sqrt[3]{4a^2}} = \frac{\left[\sqrt[3]{(4a^2)^2}\right]\sqrt[3]{3ax}}{\left[\sqrt[3]{(4a^2)^2}\right]\sqrt[3]{(4a^2)}} = \frac{\sqrt[3]{(4a^2)^2}\sqrt[3]{3ax}}{\sqrt[3]{(4a^2)^3}}$$

Note the last result is true because of the law involving radicals which states that $\sqrt[3]{a} \times \sqrt[3]{b} = \sqrt[3]{ab}$. Also, since

$$\sqrt[3]{a^3} = \left(\sqrt[3]{a}\right)^3 = \left(a^{\frac{1}{3}}\right)^3 = a^1 = a, \sqrt[3]{\left(4a^2\right)^3} = \left(\sqrt[3]{4a^2}\right)^3 = 4a^2.$$

Hence,

$$\frac{\sqrt[3]{3ax}}{\sqrt[3]{4a^2}} = \frac{\sqrt[3]{\left(4a^2\right)^2}\sqrt[3]{3ax}}{4a^2} = \frac{\sqrt[3]{16a^4}\sqrt[3]{3ax}}{4a^2}.$$

Since $\sqrt[3]{ab} = \sqrt[3]{a}\sqrt[3]{b}, \sqrt[3]{16a^4} = \sqrt[3]{\left(8a^3\right)(2a)} = \sqrt[3]{8a^3}\sqrt[3]{2a}$

$$= \sqrt[3]{\left(2a\right)^3}\sqrt[3]{2a}.$$

Note that the last result is true because $(ab)^x = a^x b^x$; that is, $8a^3 = 2^3 a^3 = (2a)^3$. Hence:

$$\frac{\sqrt[3]{3ax}}{\sqrt[3]{4a^2}} = \frac{\sqrt[3]{\left(2a\right)^3}\sqrt[3]{2a}\sqrt[3]{3ax}}{4a^2}$$

$$= \frac{2a\,\sqrt[3]{2a}\,\sqrt[3]{3ax}}{4a^2}$$

$$= \frac{2a\sqrt[3]{(2a)(3ax)}}{4a^2}$$

$$= \frac{2a\sqrt[3]{6a^2x}}{4a^2}.$$

Therefore, $\dfrac{\sqrt[3]{3ax}}{\sqrt[3]{4a^2}} = \dfrac{\sqrt[3]{6a^2x}}{2a}$.

Quiz: Fundamental Algebraic Topics

1. What is the factorization of $x^2 + ax - 2x - 2a$?

   (A) $(x+2)(x-a)$.

   (B) $(x-2)(x+a)$.

   (C) $(x+2)(x+a)$.

   (D) $(x-2)(x-a)$.

   (E) None of the above.

2. Which of the following gives

   $$\frac{2x^3 + 6x^2 - 2x - 6}{4x+4} \div \frac{x^2 - 4x + 3}{3x+6} \times \frac{6x^2 - 24}{2x^2 - 5x + 2}$$

   in the simplest form?

   (A) $\dfrac{2x^5 - 3x^4 + 18x - 26x^2 + 9}{36x^2 - 144}$.

   (B) $\dfrac{x^5 - x^4 + x^3 - 13x^2 + 9}{x^2 - 24}$.

   (C) $\dfrac{9x^3 + 63x^2 + 144x + 108}{2x^2 - 7x + 3}$.

   (D) $\dfrac{3x^3 + 27x^2 - 18x - 108}{x^2 - 5x - 1}$.

   (E) $\dfrac{2x^3 + 3x^2 - 13}{x^2 + 5x + 3}$.

3. Given $\dfrac{(\alpha + x) + y}{x + y} = \dfrac{\beta + y}{y}$, $\dfrac{x}{y} = ?$

   (A) $\dfrac{\alpha}{\beta}$.

   (B) $\dfrac{\beta}{\alpha}$.

   (C) $\dfrac{\beta}{\alpha} - 1$.

   (D) $\dfrac{\alpha}{\beta} - 1$.

   (E) 1.

4. $\sqrt{x\sqrt{x\sqrt{x}}} = ?$

    (A) $x^{\frac{7}{8}}$.

    (B) $x^{\frac{7}{14}}$.

    (C) $x^{\frac{5}{16}}$.

    (D) $x^{\frac{3}{4}}$.

    (E) $x^{\frac{15}{8}}$.

5. $4^{x-3} = \left(\sqrt{2}\right)^{x}$ The value of x is

    (A) 0.

    (B) 5.

    (C) 4.

    (D) $\dfrac{1}{2}$.

    (E) 3.

6. If $x^{64} = 64$ then $x^{32} =$

    (A) 8.

    (B) 12 or –12.

    (C) 16.

    (D) 32 or –32.

    (E) 48.

7. If $\sqrt{x-1} = 2$ then $(x-1)^{2} =$

    (A) 4.

    (B) 6.

    (C) 8.

    (D) 10.

    (E) 16.

8. The quotient of $\dfrac{\left(x^2 - 5x + 3\right)}{\left(x + 2\right)}$ is

(A) $x - 7 + \dfrac{17}{x+2}$.

(D) $x - 3 - \dfrac{3}{x+2}$.

(B) $x - 3 \dfrac{+9}{x+2}$.

(E) $x + 3 - \dfrac{3}{x+2}$.

(C) $x - 7 \dfrac{-11}{x+2}$.

9. The fraction

$$\frac{7x - 11}{x^2 - 2x - 15}$$

was obtained by adding the two fractions $\dfrac{A}{x-5} + \dfrac{B}{x+3}$.

The values of A and B are:

(A) $A = 7, B = 11$.

(D) $A = 5, B = -3$.

(B) $A = -11, B = 7$.

(E) $A = 7, B = 11$.

(C) $A = 3, B = 4$.

10. The expression $(x + y)^2 + (x - y)^2$ equals

(A) $2x^2$.

(D) $2x^2 + y^2$.

(B) $4x^2$.

(E) $x^2 + 2y^2$.

(C) $2(x^2 + y^2)$.

ANSWER KEY

1.	(B)		6.	(A)
2.	(C)		7.	(E)
3.	(D)		8.	(A)
4.	(A)		9.	(C)
5.	(C)		10.	(C)

CHAPTER 4

Solving Equations and Inequalities

4.1 Solving Equations in One Variable

There are a number of procedures which can be used in equation solving situations. Here are generalizations which are helpful in this regard.

(1) For every real number a, for every real number b, for every real number c,

$a = b$ if and only if $a + c = b + c$

$a = b$ if and only if $ac = bc$ $(c \neq 0)$

$a = b$ if and only if $a - c = b - c$

$a = b$ if and only if $\dfrac{a}{c} = \dfrac{b}{c}$ $(c \neq 0)$

(2) If a and b are real numbers and $ab = 0$, then

$a = 0$ or $b = 0$.

Here is an example which illustrates these procedures.

EXAMPLE

Solve

$$2x^2 - x - 1 = 0$$

$$(2x + 1)(x - 1) = 0$$

$$2x + 1 = 0 \quad \text{or} \quad x - 1 = 0$$

$$x = -\frac{1}{2} \quad \text{or} \quad x = 1$$

The quadratic formula follows.

If a, b, and c are real numbers, $a \neq 0$, and $ax^2 + bx + c = 0$, then

$$x = \frac{-b \pm \sqrt{b^2 - 4ac}}{2a}.$$

The following example illustrates the application of this formula.

EXAMPLE

Solve

$$x^2 + 3x - 2 = 0$$

$$x = \frac{-3 \pm \sqrt{9 - 4 \times 1(-2)}}{2}$$

$$= \frac{-3 \pm \sqrt{17}}{2}$$

Problem Solving Example:

 Solve the equation $\dfrac{2x^2 + 5x + 3}{x^2 - 2x - 8} = \dfrac{2x + 1}{x - 1}$

 We shall begin by eliminating the fractions by multiplying both sides of the equation by the least common denominator.

Thus, we now have

$$2x^3 + 3x^2 - 2x - 3 = 2x^3 - 3x^2 - 18x - 8.$$

Now, rewrite the equation so that we have all non-zero terms on the left hand side of the equal sign and zero on the right hand side of the equal sign.

$$6x^2 + 16x + 5 = 0$$

We are now able to use the quadratic formula.

$$x = \frac{-16 \pm \sqrt{(16)^2 - 4(6)(5)}}{2(6)}$$

$$= \frac{-16 \pm \sqrt{256 - 120}}{12}$$

$$= \frac{-16 \pm \sqrt{136}}{12}$$

$$= \frac{-8 \pm \sqrt{34}}{6}$$

4.2 Solving Inequalities

Here are some generalizations about inequalities which are helpful in the solution of inequalities.

(1) For every real number a, for every real number b, for every real number c,

$a < b$ if and only if $a + c < b + c$

$a < b$ if and only if $a - c < b - c$

(2) For every real number a, for every real number b, for every real number c, $c > 0$,

$a < b$ if and only if $ac < bc$

$a < b$ if and only if $\dfrac{a}{c} < \dfrac{b}{c}$

(3) For every real number a, for every real number b, for every real number c, $c < 0$,

$a < b$ if and only if $ac > bc$

$a < b$ if and only if $\dfrac{a}{c} > \dfrac{b}{c}$

Comparable generalizations could be made about ">." More specifically, if in the generalizations above, every "less than" symbol is replaced by a "greater than" symbol and every "greater than" symbol is replaced by a "less than" symbol, appropriate generalizations will result. Here is an example which illustrates the use of these generalizations.

EXAMPLE

Solve

$$-2x + 3 < 8$$

$$(-2x + 3) + (-3) < 8 + (-3)$$

$$-2x < 5$$

$$(-2x)\left(-\frac{1}{2}\right) > 5\left(-\frac{1}{2}\right)$$

$$x > -\frac{5}{2}$$

Notice that in the next to the last step, when both sides of the inequality are multiplied by a negative number, the inequality changes from "less than" to "greater than."

Problem Solving Examples:

  Solve the inequality $2x + 5 > 9$.

  $2x + 5 + (-5) > 9 + (-5)$, adding -5 to both sides.

$2x + 0 > 9 + (-5)$, additive inverse property.

$2x > 9 + (-5)$, additive identity property.

$2x > 4$, combining terms.

$\frac{1}{2}(2x) > \frac{1}{2} \times 4$, multiplying both sides by $\frac{1}{2}$.

$x > 2$

The solution set is

$$X = \left\{ x \mid 2x + 5 > 9 \right\}$$
$$= \left\{ x \mid x > 2 \right\}$$

(that is, all x such that x is greater than 2).

 Solve the inequality $4x + 3 < 6x + 8$.

 In order to solve the inequality $4x + 3 < 6x + 8$, we must find all values of x which make it true. Thus, we wish to obtain x alone on one side of the inequality.

Add -3 to both sides:

$$\begin{array}{r} 4x + 3 < 6x + 8 \\ \underline{-3 \qquad -3} \\ 4x < 6x + 5 \end{array}$$

Add $-6x$ to both sides:

$$\begin{array}{r} 4x < 6x + 5 \\ \underline{-6x \quad -6x} \\ -2x < 5 \end{array}$$

In order to obtain x alone we must divide both sides by (-2). Recall that dividing an inequality by a negative number reverses the inequality sign, hence

$$\frac{-2x}{-2} > \frac{5}{-2}$$

Cancelling $\dfrac{-2}{-2}$ we obtain, $x > -\dfrac{5}{2}$.

Thus, our solution is $\left\{ x \mid x > -\dfrac{5}{2} \right\}$ (the set of all x such that x is greater than $-\dfrac{5}{2}$).

 Find the solution set of inequality $5x - 9 > 2x + 3$.

 To find the solution set of the inequality $5x - 9 > 2x + 3$, we wish to obtain an equivalent inequality in which each term in one member involves variables, and each term in the other member is a constant. Thus, if we add $(-2x)$ to both members, only one side of the inequality will have an x term:

$$5x - 9 + (-2x) > 2x + 3 + (-2x)$$
$$5x + (-2x) - 9 > 2x + (-2x) + 3$$
$$3x - 9 > 3$$

Now, adding 9 to both sides of the inequality we obtain,

$$3x - 9 + 9 > 3 + 9$$
$$3x > 12$$

Dividing both sides by 3, we arrive at $x > 4$.

Hence the solution set is $\left\{ x \mid x > 4 \right\}$, and is pictured in the accompanying figure.

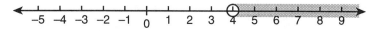

4.3   Solving Absolute Value Equations and Inequalities

Absolute value concepts occur often in problems in pre-calculus mathematics and also in calculus. The definition of absolute value follows.

For every real number x,

$$|x| = \begin{cases} x & \text{if } x \geq 0 \\ -x & \text{if } x < 0 \end{cases}$$

Here are the three generalizations which are valuable in the solution of equations and inequalities involving absolute value concepts.

For every real number a, $a > 0$,

$|x| = a$ if and only if $x = a$ or $x = -a$

$|x| < a$ if and only if $-a < x < a$

$|x| > a$ if and only if $x > a$ or $x < -a$

Here are three examples which illustrate appropriate use of these generalizations.

EXAMPLE

Solve

$|x| = 5$

$x = 5$ or $x = -5$

Solve

$|x - 2| < 3$

$-3 < x - 2 < 3$

$-1 < x < 5$

Solve

$|x + 3| > 4$

$$x + 3 > 4 \quad \text{or} \quad x + 3 < -4$$

$$x > 1 \quad \text{or} \quad x < -7$$

In the generalizations above, the requirement that $a > 0$ is significant. When that is not the case, intuitive methods can be used. For example, $|x| < -4$ has no solution, while $|x| > -4$ has every real number as a solution.

Problem Solving Examples:

 Solve for x when $|5 - 3x| = -2$.

 This problem has no solution, since the absolute value can never be negative and we need not proceed further.

 Solve for x in $|2x - 6| = |4 - 5x|$.

There are four possibilities here. $2x - 6$ and $4 - 5x$ can be either positive or negative. Therefore,

$$2x - 6 = 4 - 5x \qquad (1)$$

$$-(2x - 6) = 4 - 5x \qquad (2)$$

and

$$2x - 6 = -(4 - 5x) \qquad (3)$$

$$-(2x - 6) = -(4 - 5x) \qquad (4)$$

Equations (2) and (3) result in the same solution, as do equations (1) and (4). Therefore it is necessary to solve only for equations (1) and (2). This gives:

$$x = \frac{10}{7}, -\frac{2}{3}.$$

 Solve the inequality $|5 - 2x| > 3$.

A By the third generalization on p. 79, $|x| > a$ if and only if $x > a$ or $x < -a$. Here, x corresponds to $5 - 2x$ and a corresponds to 3. So, $5 - 2x > 3$ or $5 - 2x < -3$.

Now, we must solve for x in both inequalities. For the first, we subtract 5 from both sides of the inequality, and then divide by -2. We must keep in mind that division or multiplication by a negative number reverses the inequality sign. Thus, for $5 - 2x > 3$ we have:

$$5 - 5 - 2x > 3 - 5$$
$$-2x > -2$$
$$\frac{-2x}{-2} < \frac{-2}{-2}$$
$$x < 1.$$

For the second inequality, we subtract 5 from each side to obtain:

$$5 - 2x - 5 < -3 - 5$$
$$-2x < -8$$

Dividing by -2 yields:

$$\frac{-2x}{-2} > \frac{-8}{-2}$$
$$x > 4.$$

Therefore, the above inequality holds when $x < 1$, and when $x > 4$.

Q Solve $|3x - 1| \le 8$.

A The second generalization on p. 78 states: $|x| < a$ if and only if $-a < x < a$.

Here, x corresponds to $3x - 1$, and a corresponds to 8.

So, $-8 \le 3x - 1 \le 8$.

We must solve two equations

$$3x - 1 \le 8, \quad \text{and} \quad 3x - 1 \ge -8.$$

The solution set will be the intersection of the solution sets of each equation; that is,

$$\{x|3x-1\le 8\} \quad \text{and} \quad \{x|3x-1\ge -8\}.$$

We must find

$$\{x|3x-1\le 8\}\cap\{x|3x-1\ge -8\}$$

$$3x-1\le 8 \quad \text{and} \quad 3x-1\ge -8$$

$$3x\le 9 \quad \text{and} \quad 3x\ge -7$$

$$x\le 3 \quad \text{and} \quad x\ge -2\frac{1}{3}.$$

The solution set is $\left\{x\middle| -2\frac{1}{3}\le x\le 3\right\}$. See the figure.

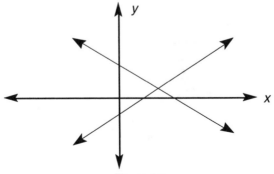

4.4 Solving Systems of Linear Equations

Consider the system of linear equations listed below.

$$a_1x + b_1y = c_1$$

$$a_2x + b_2y = c_2$$

Because the graphs of each of these equations are both straight lines, there are three geometric possibilities. One possibility is that there are two distinct and intersecting lines. This is illustrated below.

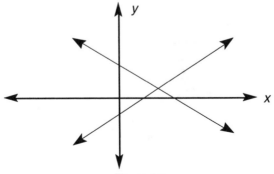

Figure 4.1

In this case, there is exactly one ordered pair which satisfies both equations. The equations are linearly independent and consistent.

A second possibility is that the two lines are parallel. This is illustrated in Figure 4.2 below.

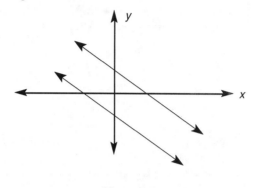

Figure 4.2

Obviously, the two lines have no points in common, so the equations have no common solution. In this case, the equations are said to be linearly independent and inconsistent.

Finally, the two equations are equations of the same line. This is illustrated in Figure 4.3 below.

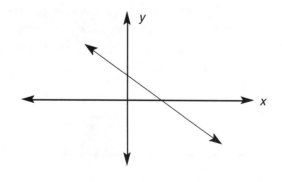

Figure 4.3

Obviously, the two equations have an infinite number of common solutions. More specifically, any solution of either equation is a solution of the other equation. In this case, the equations are said to be linearly dependent and consistent.

There are several well-known techniques for solving systems of linear equations, three of which will be illustrated here. One of the most interesting techniques for solving systems of linear equations is called Cramer's Rule. This rule involves the evaluation of determinants (determinants are reviewed in Section 9.2).

4.4.1 Cramer's Rule (In the Case of Two Equations)

Concerning the following two linear equations,

$$a_1x + b_1y = c_1$$
$$a_2x + b_2y = c_2$$

If

$$D = \begin{vmatrix} a_1 & b_1 \\ a_2 & b_2 \end{vmatrix}, \quad D_x = \begin{vmatrix} c_1 & b_1 \\ c_2 & b_2 \end{vmatrix}, \quad \text{and} \quad D_y = \begin{vmatrix} a_1 & c_1 \\ a_2 & c_2 \end{vmatrix}$$

and $D \neq 0$, then

$$x = \frac{D_x}{D} \quad \text{and} \quad y = \frac{D_y}{D}.$$

When $D = 0$, $D_x \neq 0$, and $D_y \neq 0$, the system of equations is linearly independent and inconsistent, while, when $D = D_x = D_y = 0$, the system of equations is said to be linearly dependent and consistent. Here is an example which illustrates the use of Cramer's Rule to solve a system of two equations in two unknowns.

EXAMPLE

Solve the following system of equations using Cramer's Rule.

$$x + y = 1$$
$$2x - y = 5$$

$$D = \begin{vmatrix} 1 & 1 \\ 2 & -1 \end{vmatrix}, \quad D_x = \begin{vmatrix} 1 & 1 \\ 5 & -1 \end{vmatrix}, \text{ and } D_y = \begin{vmatrix} 1 & 1 \\ 2 & 5 \end{vmatrix}$$

$$= -3 \qquad\qquad = -6 \qquad\qquad = 3$$

$$x = \frac{D_x}{D}, \quad y = \frac{D_y}{D}$$

$$= 2 \qquad = -1$$

Notice that the determinant D_x is obtained by replacing the x-coefficients in D by the corresponding constant terms, and D_y is obtained by replacing the y-coefficients in D by the constant terms. Cramer's Rule also applies in the case of n equations in n unknowns. More specifically, with three equations and three unknowns,

$$x = \frac{D_x}{D}, \quad y = \frac{D_y}{D}, \text{ and } z = \frac{D_z}{D}.$$

4.4.2 Inverses of Matrices

Solutions of systems of equations can be determined using the inverses of matrices. For example, the system of equations

$$a_1 x + b_1 y = c_1$$

$$a_2 x + b_2 y = c_2$$

can be represented by

$$AX = B$$

where $\quad A = \begin{bmatrix} a_1 & b_1 \\ a_2 & b_2 \end{bmatrix}, \ X = \begin{bmatrix} x \\ y \end{bmatrix}, \text{ and } B = \begin{bmatrix} c_1 \\ c_2 \end{bmatrix}.$

Then, $\quad A^{-1}(AX) = A^{-1}B \quad$ and $\quad X = A^{-1}B.$

Of course A^{-1} is the multiplicative inverse of A. Once X is found, it is easy to find x and y. This technique works for any system of n linear equations in n unknowns as long as A^{-1} exists. The finding of inverses of square matrices is reviewed in Section 9.3. Here is an example which illustrates the use of matrices to solve a system of two equations in two unknowns.

EXAMPLE

Solve the following system of equations using the inverse of a matrix.

$$x + y = 1$$
$$2x - y = 5$$

$$A = \begin{bmatrix} 1 & 1 \\ 2 & -1 \end{bmatrix}, \quad A^{-1} = \begin{bmatrix} \frac{1}{3} & \frac{1}{3} \\ \frac{2}{3} & -\frac{1}{3} \end{bmatrix}, \quad B = \begin{bmatrix} 1 \\ 5 \end{bmatrix}.$$

$$X = A^{-1}B = \begin{bmatrix} \frac{1}{3} & \frac{1}{3} \\ \frac{2}{3} & -\frac{1}{3} \end{bmatrix} \begin{bmatrix} 1 \\ 5 \end{bmatrix} = \begin{bmatrix} 2 \\ -1 \end{bmatrix}$$

$$\begin{bmatrix} x \\ y \end{bmatrix} = \begin{bmatrix} 2 \\ -1 \end{bmatrix}$$

$$x = 2 \quad \text{and} \quad y = -1$$

4.4.3 The Method of Substitution

The method of substitution uses one equation to solve for one variable in terms of the other. This variable is then substituted in the second equation to have one equation using one variable. Solve for this variable, and then use it to obtain the other variable by virtue of the first equation.

EXAMPLE

$$3x - y = 2 \qquad (1)$$
$$x - 2y = 1 \qquad (2)$$

Solve for y from equation (1) in terms of x:

$$y = 3x - 2$$

use this in equation (2):

$$x - 2(3x - 2) = 1$$

solve for x:

$$-5x = -3, \text{ so } x = \frac{3}{5};$$

then $y = 3\left(\dfrac{3}{5}\right) - 2 = -\dfrac{1}{5}$.

Problem Solving Examples:

 Solve the equations $2x + 3y = 6$ and $4x + 6y = 7$ simultaneously.

 We have two equations in two unknowns,

$$2x + 3y = 6 \tag{1}$$

and

$$4x + 6y = 7 \tag{2}$$

We illustrate a fourth method here. We multiply each equation by a different number so that when the two equations are added, one of the variables drops out. Thus

multiplying equation (1) by 2: $\qquad 4x + 6y = 12 \qquad$ (3)

multiplying equation (2) by –1: $\qquad -4x - 6y = -7 \qquad$ (4)

adding equations (3) and (4): $\qquad\qquad 0 = 5$

Actually, what we have shown in this case is that if there were a simultaneous solution to the given equations, then 0 would equal 5. But the conclusion is impossible; therefore there can be no simultaneous solution to these two equations, hence no point satisfying both.

If they never intersect, the straight lines which are the graphs of these equations must be parallel but not identical, which can be seen from the graph of these equations (see the accompanying diagram).

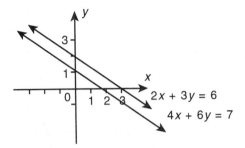

 Solve for x and y.

$$3x + 2y = 23 \tag{1}$$

$$x + y = 9 \tag{2}$$

 Multiply equation (2) by -3:

$$-3x - 3y = -27 \tag{3}$$

Add equations (1) and (3):

$$3x + 2y = 23$$
$$\underline{-3x - 3y = -27}$$
$$-y = -4$$
$$y = 4$$

Substitute 4 for y in equation (1):

$$3x + 2(4) = 23$$
$$3x + 8 = 23$$

Subtract 8 from both sides:

$$3x = 15$$

Divide each side by 3: $x = 5$

Hence our solution is, $x = 5$ and $y = 4$.

Check: Substitute 5 for x and 4 for y in equation (1):

$$3(5) + 2(4) = 23$$
$$23 = 23$$

Substitutes 5 for x and 4 for y in equation (2):

$$5 + 4 = 9$$
$$9 = 9$$

 Determine the nature of the system of linear equations

$$2x + y = 6 \tag{1}$$
$$4x + 2y = 8 \tag{2}$$

A These linear equations may be written in the standard form $y = mx + b$:

$$y = -2x + 6 \tag{3}$$
$$\text{and } y = -2x + 4 \tag{4}$$

Observe that the slope of each line is $m = -2$, but the y–intercepts are different, that is, $b = 6$ for equation (31) and $b = 4$ for equation (42). The lines are therefore parallel and distinct. The graph below also indicates that the lines are parallel. The system is therefore inconsistent, and there is no solution.

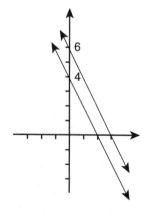

Determine the nature of the system of linear equations

$$x + 3y = 4 \tag{1}$$

$$2x + 6y = 8. \tag{2}$$

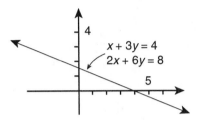

 If the first equation is multiplied by 2, the solution of the system will not be altered. Note, however, that the two equation are then identical. The graph too, indicates that the lines coincide, and therefore the system is consistent and dependent. It can be verified by substitution that three of the solutions are $x = 1$, $y = 1$; $x = 7$, $y = -1$; and, $x = -5$, $y = 3$.

Solve the system of linear equations:

$$3x + 2y + 4z = 1$$
$$2x - y + z = 0$$
$$x + 2y + 3z = 1.$$

Use Cramer's Rule to solve this system. Write the above equations in matrix form:

$$\begin{bmatrix} 3 & 2 & 4 \\ 2 & -1 & 1 \\ 1 & 2 & 3 \end{bmatrix} \begin{bmatrix} x \\ y \\ z \end{bmatrix} = \begin{bmatrix} 1 \\ 0 \\ 1 \end{bmatrix}.$$

Then

$$A = \begin{bmatrix} 3 & 2 & 4 \\ 2 & -1 & 1 \\ 1 & 2 & 3 \end{bmatrix}, \quad B = \begin{bmatrix} 1 \\ 0 \\ 1 \end{bmatrix}.$$

First, check that det $A \neq 0$.

$$\det(A) = \begin{vmatrix} 3 & 2 & 4 \\ 2 & -1 & 1 \\ 1 & 2 & 3 \end{vmatrix}$$

$$\det(A) = 3\begin{vmatrix} -1 & 1 \\ 2 & 3 \end{vmatrix} - 2\begin{vmatrix} 2 & 1 \\ 1 & 3 \end{vmatrix} + 4\begin{vmatrix} 2 & -1 \\ 1 & 2 \end{vmatrix}$$

$$= 3(-3-2) - 2(6-1) + 4(4+1)$$

$$= -15 - 10 + 20 = -5.$$

Since det $A \neq 0$, the system has a unique solution. Then,

$$x = \frac{\det(A_1)}{\det(A)}, y = \frac{\det(A_2)}{\det(A)}, z = \frac{\det(A_3)}{\det(A)}.$$

$\det(A_1)$ is the determinant of the matrix obtained by replacing the first column of A by the column vector B.

Thus,

$$\det(A_1) = \begin{bmatrix} 1 & 2 & 4 \\ 0 & -1 & 1 \\ 1 & 2 & 3 \end{bmatrix}.$$

Expand the determinant by minors, using the first column.

$$\det(A_1) = 1\begin{vmatrix} -1 & 1 \\ 2 & 3 \end{vmatrix} + 1\begin{vmatrix} 2 & 4 \\ -1 & 1 \end{vmatrix} =$$

$$= 1(-3-2) + 1(2+4)$$

$$= -5 + 6 = +1.$$

Now, we have $\det(A) = -5.$

Thus,

$$x = \frac{\det(A_1)}{\det(A)} = \frac{1}{-5} = -\frac{1}{5}.$$

$$y = \frac{\det(A_2)}{\det(A)} = \frac{\begin{vmatrix} 3 & 1 & 4 \\ 2 & 0 & 1 \\ 1 & 1 & 3 \end{vmatrix}}{-5}.$$

Now, expand $\det(A_2)$ along the second row:

$$\begin{vmatrix} 3 & 1 & 4 \\ 2 & 0 & 1 \\ 1 & 1 & 3 \end{vmatrix} = -2 \begin{vmatrix} 1 & 4 \\ 1 & 3 \end{vmatrix} - 1 \begin{vmatrix} 3 & 1 \\ 1 & 1 \end{vmatrix}$$

$$= -2(3-4) - 1(3-1)$$

$$= 2 - 2 = 0$$

$$y = \frac{0}{-5} = 0.$$

$$z = \frac{\det(A_3)}{\det(A)} = \frac{\begin{vmatrix} 3 & 2 & 1 \\ 2 & -1 & 0 \\ 1 & 2 & 1 \end{vmatrix}}{-5}$$

Expand determinant A_3 by minors, using the third column.

$$\begin{vmatrix} 3 & 2 & 1 \\ 2 & -1 & 0 \\ 1 & 2 & 1 \end{vmatrix} = +1 \begin{vmatrix} 2 & -1 \\ 1 & 2 \end{vmatrix} + 1 \begin{vmatrix} 3 & 2 \\ 2 & -1 \end{vmatrix}$$

$$= (4+1) + (-3-4)$$

$$= 5 - 7 = -2.$$

Then,

$$z = \frac{-2}{-5} = \frac{2}{5}.$$

Thus $x = -\frac{1}{5}, y = 0, z = \frac{2}{5}.$

4.5 Solving Systems of Linear Inequalities and Linear Programming

In the previous section, several approaches were given to the solution of systems of linear equations. In this section, systems of linear inequalities will be examined from a geometric viewpoint. For example, for specific real numbers a, b, and c, the graph of

$$ax + by + c = 0$$

is a line, while the graph of

$$ax + by + c < 0$$

is the half plane on one side of the line, with the graph of

$$ax + by + c > 0$$

being the half plane on the other side of the line. Consider the system of inequalities below.

$$x + y \leq 6$$

$$x - y \geq -2$$

$$x \geq 0$$

$$y \geq 0$$

The solutions for this system are the set of all points in the shaded region (including the boundary points).

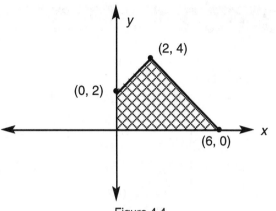

Figure 4.4

In a linear programming problem, an objective function

$$C = ax + by + c$$

is given subject to constraints specified by linear inequalities. Then, if C assumes a maximum (or minimum), this maximum (minimum) must occur at a vertex for the system of linear inequalities. Suppose

$$C = 2x + y$$

and the objective is to maximize or minimize C subject to the system of inequalities given above. A table is helpful.

Vertex	Corresponding value of C
(0, 0)	$2 \times 0 + 0 = 0$
(0, 2)	$2 \times 0 + 2 = 2$
(2, 4)	$2 \times 2 + 4 = 8$
(6, 0)	$2 \times 6 + 0 = 12$

Table 4.1

Thus, the maximum value for C occurs at (6,0) while the minimum value occurs at (0,0).

Problem Solving Examples:

 Graph the following inequalities and indicate the region of their intersection:

$$x \geq 1, \quad y \geq 0, \quad x + y \leq 4, \quad 4x + 3y \leq 14.$$

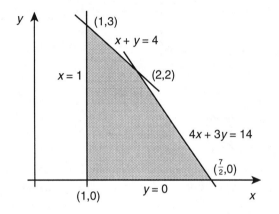

 The solution is the shaded region, having the quadrilateral as its boundary. The indicated vertices are the intersection points of the inequalities.

 Solve the following system graphically.

$$y - x > -3$$
$$y - 2x < 2$$
$$x + y - 3 < 0$$

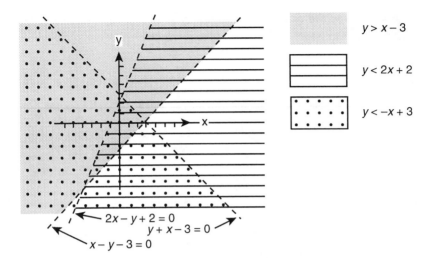

$y > x - 3$

$y < 2x + 2$

$y < -x + 3$

 We may rewrite the system:

$$y > x - 3$$
$$y < 2x + 2$$
$$y < -x + 3$$

Graph the linear equation, $y = mx + b$, for each inequality as a straight dotted line. Thus, we graph

$$y = x - 3$$
$$y = 2x + 2$$
$$y = -x + 3$$

To determine in what region of the x–y plane the inequality holds, select points on both sides of the corresponding dotted line and substitute them into the variable statement. Shade in the side of the line whose point makes the inequality a true statement.

The graphs of the variable sentences are represented in the accompanying figure by solid, horizontal, and dotted shading, respectively.

The triple-shaded triangular region is the set of all points whose coordinate pairs satisfy all three conditions as defined by the three inequalities in the system.

Q Draw the graph of the given system of inequalities, and determine the coordinates of the vertices of the polygon which forms the boundary.

$$y \le 3x - 3 \tag{1}$$

$$3y \le 24 - 2x \tag{2}$$

$$2y \ge 3x - 10 \tag{3}$$

$$y \ge -x + 5 \tag{4}$$

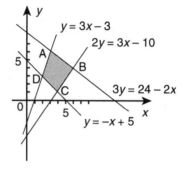

A y is expressed in terms of x. For each inequality draw the corresponding equality. Choose a point on each side of each solid line to determine the area where the particular inequality holds. Shade in that region. The graph of the given system of inequalities consists of the hatched area and the four lines which form the boundary, that is, the polygon $ABCD$. The vertex A is found by solving the system obtained by writing equations corresponding to inequalities (1) and (2):

$$y = 3x - 3 \tag{5}$$

$$3y = 24 - 2x \tag{6}$$

Solving the system of equations (5) and (6), we have

$$x = 3, \qquad y = 6$$

The coordinates of the vertex A are therefore $(3, 6)$. In a similar manner, the coordinates of B, C, and D are found to be $(6, 4)$ $(4, 1)$ and $(2, 3)$, respectively.

4.6 Solving Equations Involving Radicals, Solving Rational Equations, and Solving Equations in Quadratic Form

In solving equations involving radicals, it is sometimes necessary to raise both sides of an equation to the same positive integral power. In this regard, the following generalization is applied:

If $P(x)$ and $Q(x)$ are expressions in x, and n is a positive integer, then the set of all solutions of the equation $P(x) = Q(x)$ is a subset of the set of all solutions of

$$[P(x)]^n = [Q(x)]^n.$$

This means that when both sides of an equation are raised to the same power, the resulting equation may have solutions which are not solutions of the original equation. Thus, it is necessary to check to see if solutions of the new equation are actually solutions of the original equation. This is illustrated in the example below.

EXAMPLE

Solve

$$\sqrt{x+2} - x = -4$$

$$\sqrt{x+2} = x - 4$$
$$(\sqrt{x+2})^2 = (x-4)^2$$
$$x + 2 = x^2 - 8x + 16$$
$$0 = x^2 - 9x + 14$$
$$0 = (x-7)(x-2)$$
$$x = 7 \quad \text{or} \quad x = 2$$
$$\sqrt{7+2} - 7 = -4 \quad \text{but}$$
$$\sqrt{2+2} - 2 \neq -4$$

Thus, $x = 7$ is the only solution.

Equations often contain fractions that have denominators which are polynomials. Such equations are called rational equations. In solving equations of this type, it is often desirable to multiply both sides of the equation by a polynomial to eliminate all fractions. This must be done cautiously. Here is a generalization which was stated originally in Section 4.1.

For every real number a, for every real number b, for every real number c, $c \neq 0$.

$a = b$ if and only if $ac = bc$

The important idea is that $c \neq 0$. In practice, it is common to merely multiply both sides by the appropriate polynomial and then check at the end. Here is an example which illustrates this procedure.

EXAMPLE

Solve

$$\frac{x}{x-2} + 1 = \frac{x^2 + 4}{x^2 - 4}$$

$$\left(\frac{x}{x-2} + 1\right)(x-2)(x+2) = \frac{x^2+4}{x^2-4}(x+2)(x-2)$$

$$x(x+2) + (x-2)(x+2) = x^2 + 4$$

$$2x^2 + 2x - 4 = x^2 + 4$$

$$x^2 + 2x - 8 = 0$$

$$(x+4)(x-2) = 0$$

$$x + 4 = 0 \quad \text{or} \quad x - 2 = 0$$

$$x = -4 \quad \text{or} \quad x = 2$$

When $x = -4$,

$$\frac{-4}{-4-2} + 1 = \frac{(-4)^2 + 4}{(-4)^2 - 4}$$

but when $x = 2$

$$\frac{2}{2-2}+1 \neq \frac{2^2+4}{2^2-4}, \text{ with zero denominators.}$$

Thus, the only solution is $x = -4$.

Equations of the form

$$ax^{2n} + bx^n + c = 0$$

are said to be in quadratic form. Equations of this type can often be solved by an appropriate substitution, as illustrated below.

EXAMPLE

Solve

$$6x^{-2} - x^{-1} - 2 = 0$$

Let $y = x^{-1}$

$$6y^2 - y - 2 = 0$$

$$(3y - 2)(2y + 1) = 0$$

$$3y - 2 = 0 \quad \text{or} \quad 2y + 1 = 0$$

$$y = \frac{2}{3} \qquad \text{or} \qquad y = -\frac{1}{2}$$

$$x^{-1} = \frac{2}{3} \qquad \text{or} \qquad x^{-1} = -\frac{1}{2}$$

$$x = \frac{3}{2} \qquad \text{or} \qquad x = -2$$

Problem Solving Examples:

 Solve the equation $\sqrt{2x^2 - 9} = x$.

 Squaring both sides, we have:

$$2x^2 - 9 = x^2$$

$$x^2 = 9$$

$$x = 3 \quad \text{or} \quad x = -3$$

Both 3 and –3 will satisfy the equation $2x^2 - 9 = x^2$ since $2(3)^2 - 9 = 9 = (3)^2$ and $2(-3)^2 - 9 = 9 = (-3)^2$. However, –3 does not satisfy the original equation since $\sqrt{2(-3)^2 - 9} = \sqrt{9} = 3 \neq -3$. An extraneous root was introduced by squaring. Thus the solution set is $\{3\}$.

 Solve $\dfrac{x}{x-2} + \dfrac{x-1}{2} = x+1$.

 First eliminate the fractions to facilitate the solution. This is done by multiplying both sides of the equation by the Least Common Denominator, LCD. The LCD is obtained by multiplying the denominators of every fraction: LCD = $2(x - 2)$. Multiplying each side by this, the equation becomes:

$$2(x-2)\left(\frac{x}{x-2} + \frac{x-1}{2}\right) = (x+1)2(x-2)$$
$$2x + (x-1)(x-2) = 2(x+1)(x-2)$$
$$2x + x^2 - 3x + 2 = 2x^2 - 2x - 4$$
$$x^2 - x - 6 = 0$$

This can be solved by factoring and setting each factor equal to zero.

$$(x-3)(x+2) = 0$$
$$x - 3 = 0 \quad x + 2 = 0$$
$$x = 3 \qquad x = -2$$

Since both of these solutions are admissible values of x, they both should satisfy the original equation.

Check for $x = 3$:

$$\frac{3}{1} + \frac{2}{2} = 3 + 1$$
$$3 + 1 = 3 + 1$$

Check for $x = -2$:

$$\frac{-2}{-4} + \frac{-3}{2} = -2 + 1$$

$$\frac{1}{2} + \frac{-3}{2} = -1$$

$$-1 = -1$$

Solve the equation $x^4 - 5x^2 - 36 = 0$.

This is a fourth degree equation, but it can be solved by the same methods applied to quadratic equations.

To solve $x^4 - 5x^2 - 36 = 0$, we let $z = x^2$, substitute in the given equation, and get

$$z^2 - 5z - 36 = 0$$

This is now a quadratic equation in the variable z. We solve this equation by factoring.

$$z^2 - 5z - 36 = 0$$

Factoring:

$$(z - 9)(z + 4) = 0$$

$z - 9 = 0, \quad z + 4 = 0$, setting both factors equal to zero.

$z = 9, \quad z = -4$, solving for z.

Hence the solution set of the equation in z is $\{-4, 9\}$. Now we replace z in $z = x^2$ by -4 and then by 9 and get

$$x^2 = -4$$

Taking the square root of each member,

$$x = \pm\sqrt{-4} = \pm\sqrt{4(-1)} = \pm\sqrt{4}\sqrt{-1}$$

$$x = \pm 2i$$

Also $x^2 = 9$

$$x = \pm 3$$

Consequently the solution set of the original equation is $\{2i, -2i\} \cup \{3, -3\} = \{2i, -2i, 3, -3\}$.

 Solve $x^4 - 2x^2 - 3 = 0$ as a quadratic in x^2.

 We can write the equation as $\left(x^2\right)^2 - 2x^2 - 3 = 0$. Let $x^2 = z$ and we have

$$z^2 - 2z - 3 = 0,$$

which is a quadratic equation. We can solve a quadratic equation in the form $ax^2 + bx + c = 0$ using the quadratic formula,

$$x = \frac{-b \pm \sqrt{b^2 - 4ac}}{2a}.$$ (Note: we could also solve by factoring.)

In our case $a = 1$, $b = -2$, and $c = -3$. Thus,

$$z = \frac{-(-2) \pm \sqrt{(-2)^2 - 4(1)(-3)}}{2(1)}$$

$$= \frac{2 \pm \sqrt{4 + 12}}{2} = \frac{2 \pm 4}{2} = 1 \pm 2.$$

Therefore, $z = 3$ or $z = -1$. Since $z = x^2$, we have $x^2 = 3$ or $x^2 = -1$. If $x^2 = 3$, $x = \pm\sqrt{3}$. If $x^2 = -1$, $x = \pm i$. Hence, the solution set of the original equation is

$$\left\{\sqrt{3}, -\sqrt{3}, i, -i\right\}.$$

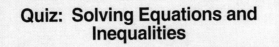

1. Solve for z:

    $7z + 1 - z = 2z - 7$

    (A) −2.   (D) 2.

    (B) 0.   (E) 3.

    (C) 1.

2. Solve for p:

    $0.4p + 1 = 0.7p - 2$

    (A) 0.1.   (D) 10.

    (B) 2.   (E) 12.

    (C) 5.

3. Find the solution set for the pair of equations.

    $$4x + 3y = 9$$
    $$2x - 2y = 8$$

    (A) (−3,1).   (D) (3,−1).

    (B) (1,−3).   (E) (−1,3).

    (C) (3,1).

4. Solve for x: $\sqrt{x + 2} + 1 = \sqrt{x + 9}$

    (A) 3.   (D) 1.

    (B) 7.   (E) 5.

    (C) 6.

5. Solve for all values of x.

$$x^2 - 7x = -10$$

   (A) −3 or 5.   (D) −2 or −5.

   (B) 2 or 5.   (E) 5.

   (C) 2.

6. $\left|3 - \dfrac{1}{2}y\right| = -7$

   (A) −8 or 20.   (D) 4 or −2.

   (B) 8 or −20.   (E) No solution.

   (C) 2 or − 5.

7. $2|x + 7| = 12$

   (A) −13 or −1.   (D) 6 or −13.

   (B) −6 or 6.   (E) No solution.

   (C) −1 or 13.

8. Find the solution set for the inequality

$$3m + 2 < 7$$

   (A) $m \geq \dfrac{5}{3}$.   (D) $m > 2$.

   (B) $m \leq 2$.   (E) $m < \dfrac{5}{3}$.

   (C) $m < 2$.

9. The following represents the solution set for which system of inequalities?

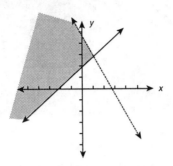

(A) $\begin{cases} y \le x+2 \\ y > -2x+5 \end{cases}$

(D) $\begin{cases} y > x+2 \\ y \le 2x+5 \end{cases}$

(B) $\begin{cases} y > x+2 \\ y < -2x+5 \end{cases}$

(E) $\begin{cases} y \ge -2x+5 \\ y < x+2 \end{cases}$

(C) $\begin{cases} y \ge x+2 \\ y < -2x+5 \end{cases}$

10. $-6 < \dfrac{2}{3}r + 6 < 2$

(A) $-6 < r < -3$.

(D) $-2 < r < -\dfrac{4}{3}$.

(B) $-18 < r < -6$.

(E) $r < -6$.

(C) $r > -6$.

ANSWER KEY

1.	(A)		6.	(E)
2.	(D)		7.	(A)
3.	(D)		8.	(E)
4.	(B)		9.	(C)
5.	(B)		10.	(B)

CHAPTER 5

Functions

5.1 Functions and Equations

There are numerous examples in mathematics where a second quantity is found from a first quantity by applying some rule. Consider the equation

$$y = 2x + 3.$$

In this case, each y-value is determined from the x-value by multiplying the x-value by 2 and then adding 3 to the result. This is an example of a function. In general, a function is a rule which specifies an output number for each input number. Usually, the input numbers are specified as x-values and the output numbers are specified as y-values. Here is the definition of the term function.

Let A and B be nonempty sets. Then a function from A to B is a rule which assigns to each element of A exactly one element of B. The domain of the function is A and the range consists only of those elements of B which are actually paired with elements of A.

Pictures are often used to illustrate functions, and the symbol f is often used to represent a function. For the function f pictured on the next page (Figure 5.1), the domain is A and

$$A = \{a, b, c\},$$

Figure 5.1

while $B = \{d, e, f\}$, and the range is $\{d, e\}$. Notice that each element of A is paired with exactly one element of B.

Many functions are merely illustrated in equation form. For example, $y = 2x + 3$ falls into that category. Unless it is specifically stated otherwise, the domain of this function is the set of all real numbers. This same function can be denoted by

$$f(x) = 2x + 3.$$

Also, $f(1) = 2 \times 1 + 3$

$$= 5.$$

This means that the number associated with 1 is 5 or, equivalently, $y = 5$ when $x = 1$.

When the function is denoted by an equation or by using the $f(x)$ notation, the domain is agreed to be the largest subset of real numbers for which y or $f(x)$ is a real number. For example, the domain of

$$f(x) = \frac{x}{x-1}$$

is the set of all real numbers except 1, while the domain of $y = \sqrt{x}$ is the set of all nonnegative real numbers.

It is quite easy to tell whether or not a given graph is the graph of a function. This is determined by the famous vertical line test. Consider the two graphs that follow.

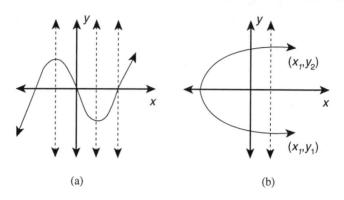

(a)   (b)

Figure 5.2

Notice that for the graph on the left, each vertical line intersects the graph in exactly one point. Thus, this is the graph of a function. On the other hand, for the graph on the right, there is a vertical line which intersects the curve in more than one point. In this case for a given first component x_1, there are two second components y_1 and y_2 and this is not the graph of a function. Here is a summary of the results described above.

(1)   When each vertical line intersects the graph in, at most, one point, the graph describes a function.

(2)   When there exists a vertical line which intersects the graph in more than one point, the graph does not describe a function.

A function is sometimes thought of as a set of points $\{(x, f(x)) | x \in \text{Domain} f\}$. This notion is used in some of the problems below.

Problem Solving Examples:

 If $f(x) = (x - 2)/(x + 1)$, find the function values $f(2)$, $f\left(\dfrac{1}{2}\right)$, and $f\left(-\dfrac{3}{4}\right)$.

 To find $f(2)$, we replace x by 2 in the given formula for $f(x)$, $f(x) = (x - 2)/(x + 1)$; thus

$$f(2) = \frac{2-2}{2+1} = \frac{0}{3} = 0.$$

Similarly, $f\left(\dfrac{1}{2}\right) = \dfrac{\dfrac{1}{2} - 2}{\dfrac{1}{2} + 1}.$

Multiply numerator and denominator by 2,

$$= \frac{2\left(\dfrac{1}{2} - 2\right)}{2\left(\dfrac{1}{2} + 1\right)}.$$

Distribute,

$$= \frac{2\left(\dfrac{1}{2}\right) - 2 \times 2}{2\left(\dfrac{1}{2}\right) + 2}$$

$$= \frac{1-4}{1+2}$$

$$= -\frac{3}{3} = -1.$$

$$f\left(-\frac{3}{4}\right) = \frac{-\dfrac{3}{4} - 2}{-\dfrac{3}{4} + 1}.$$

Multiply numerator and denominator by 4,

$$= \frac{4\left(-\dfrac{3}{4} - 2\right)}{4\left(-\dfrac{3}{4} + 1\right)}.$$

Distribute,

$$= \frac{4\left(-\dfrac{3}{4}\right) - 4(2)}{4\left(-\dfrac{3}{4}\right) + 4(1)}$$

$$= \frac{-3 - 8}{-3 + 4}$$

$$= \frac{-11}{1}$$

$$= -11.$$

If $g(x) = x^2 - 2x + 1$, find the given element in the range.

(a) $g(-2)$

(b) $g(0)$

(c) $g(a + 1)$

(d) $g(a - 1)$

(a) To find $g(-2)$, substitute -2 for x in the given equation.

$$\begin{aligned} g(x) &= g(-2) \\ &= (-2)^2 - 2(-2) + 1 \\ &= 4 + 4 + 1 \\ &= 8 + 1 \\ &= 9 \end{aligned}$$

Therefore, $g(-2) = 9$.

(b) To find $g(0)$, substitute 0 for x in the given equation.

$$\begin{aligned} g(x) &= g(0) \\ &= (0)^2 - 2(0) + 1 \\ &= 0 - 0 + 1 \\ &= 1 \end{aligned}$$

Hence, $g(0) = 1$.

(c) To find $g(a+1)$, substitute $a + 1$ for x in given equation.

$$g(x) = g(a+1)$$
$$= (a+1)^2 - 2(a+1) + 1$$
$$= (a^2 + 2a + 1) - 2a - 2 + 1$$
$$= a^2 + 2a + 1 - 2a - 2 + 1$$
$$= a^2 + 1 - 2 + 1$$
$$= a^2 + 0$$
$$= a^2.$$

Therefore, $g(a+1) = a^2$.

(d) To find $g(a-1)$, substitute $a - 1$ for x in given equation.

$$g(x) = g(a-1)$$
$$= (a-1)^2 - 2(a-1) + 1$$
$$= (a^2 - 2a + 1) - 2a + 2 + 1$$
$$= a^2 - 2a + 1 - 2a + 2 + 1$$
$$= a^2 - 4a + 4$$

Hence, $g(a-1) = a^2 - 4a + 4$.

 Find the domain D and the range R of the function $\left(x, \dfrac{x}{|x|} \right)$.

Note that the y-value of any coordinate pair (x,y) is $\dfrac{x}{|x|}$.

We can replace x in the formula $\dfrac{x}{|x|}$ with any number except

0, since the denominator, $|x|$, cannot equal 0, (i.e. $|x| \neq 0$) which is equivalent to $x \neq 0$. This is because division by 0 is undefined. Therefore, the domain D is the set of all real numbers except 0. If x is negative, i.e. $x < 0$, then $|x| = -x$ by definition.

Hence, if x is negative, then $\dfrac{x}{|x|} = \dfrac{x}{-x} = -1$. If x is positive, i.e.

$x > 0$, then $|x| = x$ by definition. Hence, if x is positive, then

$\dfrac{x}{|x|} = \dfrac{x}{x} = 1$. (The case where $x = 0$ has already been found to be

undefined). Thus, there are only two numbers, -1 and 1, in the range R of the function; that is, R = $\{-1,1\}$.

 Describe the domain and range of the function

$$f = \left\{(x,y) \Big| y = \sqrt{9-x^2}\right\} \text{ if } x \text{ and } y \text{ are real numbers.}$$

In determining the domain we are interested in the values of x which yield a real value for y. Since the square root of a negative number is not a real number, the domain is restricted to those values of x which make the radicand positive or zero. Therefore x^2 cannot exceed 9, which means that x cannot exceed 3 or be less that -3. A convenient way to express this is to write $-3 \le x \le 3$, which is read "x is greater than or equal to -3 and less than or equal to 3." This is the domain of the function. The range is the set of values that y can assume. To determine the range of the function we note that the largest value of y occurs when $x = 0$. Then $y = \sqrt{9-0} = 3$. Likewise, the smallest value of y occurs when $x = 3$ or $x = -3$. Then $y = \sqrt{9-9} = 0$. Since this is an inclusive interval of the real axis, the range of y is $0 \le y \le 3$.

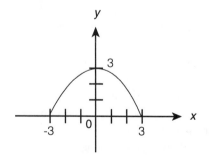

Given the following graphs:
Indicate (a) which graph represents a binary relation?
(b) which are graphs of functions?

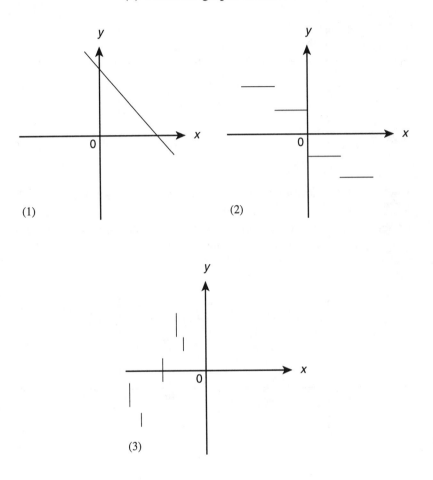

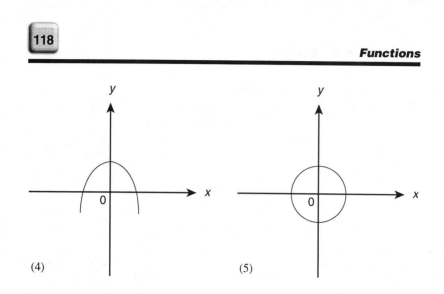

(4)   (5)

A   (a) By definition, a binary relation in A is a set of ordered pairs whose first and second coordinates are both members of set A. Here, A is the set of real numbers. Therefore, all the given graphs are graphs of binary relations.

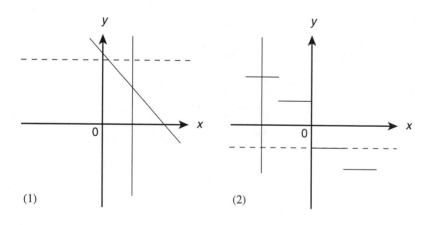

(1)   (2)

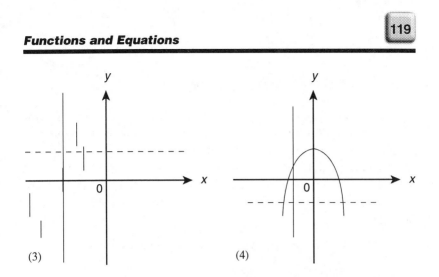

(3)   (4)

(b) A function f from set A into set B is defined as a set of ordered pairs $\{(a,b)|a \in A$ and $b \in B\}$ no two of which have the same first coordinate. Hence, by the vertical-line test (see figure), it's found that graphs (1), (2) and (4) are graphs of functions from X to Y where X

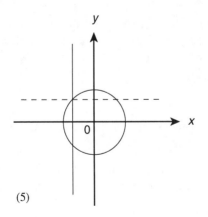

(5)

and Y are set of x-coordinates and y-coordinates respectively. Graphs (1) and (3) are graphs of functions from Y to X by the horizontal-line test (see figure).

5.2   Combining Functions

There are various ways in which two real numbers can be combined to form a third real number. For example, the sum, difference, product, and quotient of two real numbers can be found. Similarly, two functions can be combined to form a third function, and the most natural way to accomplish this is to do it in the context of sum, difference, product, and quotient. The definitions of these function operations follow.

Let f and g be functions with a common domain. Then,

$$(f + g)(x) = f(x) + g(x)$$

$$(f - g)(x) = f(x) - g(x)$$

$$(fg)(x) = [f(x)][g(x)]$$

and $\left(\dfrac{f}{g}\right)(x) = \dfrac{f(x)}{g(x)}, \quad g(x) \neq 0.$

Two functions can also be combined to form a third function using a process called the composition of functions. If f and g are functions, then the composition of these functions is denoted by $f \circ g$ and is defined by

$$(f \circ g)(x) = f(g(x)).$$

The domain of this function is defined to be the set of all x, such that $g(x)$ is in the domain of f.

The composition of functions is often illustrated with pictures. The following picture represents $f \circ g$ for given functions f and g.

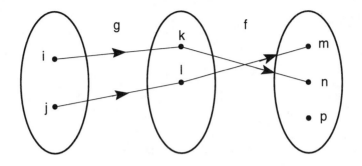

Figure 5.3

Here are some examples which show how composition and computation can be performed.

EXAMPLE

Given $f(x) = 2x - 1$ and $g(x) = x + 3$, find:

(a) $(fg)(x)$.

$$(fg)x = [f(x)][g(x)]$$
$$= (2x - 1)(x + 3)$$
$$= 2x^2 + 5x - 3$$

(b) $(f \circ g)x$.

$$(f \circ g)x = f(g(x))$$
$$= f(x + 3)$$
$$= 2(x + 3) - 1$$
$$= 2x + 5$$

(c)　$(g \circ f)x.$

$$(g \circ f)x = g(f(x))$$
$$= g(2x - 1)$$
$$= (2x - 1) + 3$$
$$= 2x + 2$$

From (b) and (c) above, it is obvious that, in general,

$$(f \circ g)(x) \neq (g \circ f)(x).$$

Problem Solving Examples:

If $f(x) = x^2 - x - 3$, $g(x) = \dfrac{\left(x^2 - 1\right)}{(x + 2)}$, and $h(x) = (f+g)(x)$, find $h(2)$.

$h(x) = f(x) + g(x)$, and we are told that $f(x) = x^2 - x - 3$ and $g(x) = \dfrac{\left(x^2 - 1\right)}{(x + 2)}$; thus $h(x) = (x^2 - x - 3) + \dfrac{\left(x^2 - 1\right)}{(x + 2)}$.

To find $h(2)$, we replace x by 2 in the above formula for $h(x)$,

$$h(2) = \left[(2)^2 - 2 - 3\right] + \left(\frac{2^2 - 1}{2 + 2}\right)$$

$$= (4 - 2 - 3) + \left(\frac{4 - 1}{4}\right)$$

$$= (-1) + \left(\frac{3}{4}\right)$$

$$= -\frac{4}{4} + \frac{3}{4}$$

$$= -\frac{1}{4}.$$

Thus,
$$h(2) = -\frac{1}{4}.$$

Let $f(x) = 2x^2$ with domain $D_f = R$ and $g(x) = x - 5$ with $D_g = R$.

Find   (a) $f + g$

       (b) $f - g$

       (c) fg

       (d) $\dfrac{f}{g}$

(a) $f + g$ has domain R and
$$(f + g)(x) = f(x) + g(x) = 2x^2 + x - 5$$

for each number x. For example,
$$(f + g)(1) = f(1) + g(1) = 2(1)^2 + 1 - 5 = 2 - 4 = -2.$$

(b) $f - g$ has domain R and
$$(f - g)(x) = f(x) - g(x) = 2x^2 - (x - 5) = 2x^2 - x + 5$$

for each number x. For example,
$$(f - g)(1) = f(1) - g(1) = 2(1)^2 - 1 + 5 = 2 + 4 = 6.$$

(c) fg has domain R and
$$(fg)(x) = f(x) \times g(x) = 2x^2 \times (x - 5) = 2x^3 - 10x^2$$

for each number x. In particular,
$$(fg)(1) = 2(1)^3 - 10(1)^2 = 2 - 10 = -8.$$

(d) $\dfrac{f}{g}$ has domain R excluding the number

$x = 5$ (when $x = 5$, $g(x) = 0$ and division by zero is undefined) and
$$\left(\frac{f}{g}\right)(x) = \frac{f(x)}{g(x)} = \frac{2x^2}{x - 5}$$

for each number $x \neq 5$. In particular,

$$\left(\frac{f}{g}\right)(1) = \frac{2(1)^2}{1-5} = \frac{2}{-4} = -\frac{1}{2}.$$

Q Let $f: R \to R$ and $g: R \to R$ be two functions, given by $f(x) = 2x + 5$ and $g(x) = 4x^2$, respectively for all x in R, where R is the set of real numbers. Here the arrow indicates that the functions take a subset of the real numbers, the domain, into a subset of the real numbers, the range. Find expressions for the compositions $(f \circ g)(x)$ and $(g \circ f)(x)$.

A Consider functions $f: A \to B$ and $g: B \to C$ — that is, where the co-domain of f is the domain of g. Then the function $g \circ f$ is defined as $g \circ f: A \to C$, where $(g \circ f)(x) = g(f(x))$ for all x in A, and it is called the composition of f and g.

Therefore,

$$(f \circ g)(x) = f(g(x))$$
$$= f(4x^2)$$
$$= 2[4x^2] + 5$$
$$= 8x^2 + 5 \quad \text{and}$$

$$(g \circ f)(x) = g(f(x))$$
$$= g(2x + 5)$$
$$= 4[2x + 5]^2$$
$$= 4(4x^2 + 20x + 25)$$
$$= 16x^2 + 80x + 100$$

It's seen from the results that $f \circ g \neq g \circ f$. This is true in general, i.e., functional compositions are not commutative operations.

5.3   Inverse Functions

Consider the functions

$$f(x) = 2x \quad \text{and} \quad g(x) = \frac{1}{2}x.$$

Then f is the "doubling function" and $g(x)$ is the "halving function," and these functions undo each other. More specifically,

$$(f \circ g)(6) = f(g(6)) \quad \text{and} \quad (g \circ f)(6) = g(f(6))$$
$$= f(3) \qquad\qquad\qquad = g(12)$$
$$= 6 \qquad\qquad\qquad\quad = 6$$

Also, $(f \circ g)(x) = x$ and $(g \circ f)(x) = x$ for each real number x. In such a case, f and g are inverses of each other. Here is the general definition of inverse functions.

> Two functions f and g are said to be inverses of each other provided that $(f \circ g)(x) = x$ for each x in the domain of g and $(g \circ f)(x) = x$ for each x in the domain of f.

The symbol $f^{-1}(x)$ is used to represent the inverse of $f(x)$. Here is a summary of a procedure for finding $f^{-1}(x)$.

(1)   Let $y = f(x)$.

(2)   Interchange x and y in $y = f(x)$.

(3)   Solve the resulting equation for y.

Here is an example which illustrates this procedure.

EXAMPLE

Given $f(x) = 2x - 3$, find $f^{-1}(x)$ and show that these two functions are actually inverse's of each other.

$$y = 2x - 3$$
$$x = 2y - 3$$
$$x + 3 = 2y$$

$$y = \frac{x+3}{2}$$

$$f^{-1}(x) = \frac{x+3}{2}$$

$(f \circ f^{-1})(x) = f(f^{-1}(x))$ $(f^{-1} \circ f)(x) = f^{-1}(f(x))$

$$= f\left(\frac{x+3}{2}\right) \qquad\qquad = f^{-1}(2x-3)$$

$$= 2\left(\frac{x+3}{2}\right) - 3 \qquad = \frac{(2x-3)+3}{2}$$

$$= x \qquad\qquad\qquad = x$$

Problem Solving Examples:

 Let f be the linear function that is defined by the equation $f(x) = 3x + 2$. Find the equation that defines the inverse function f^{-1}.

 To find the inverse function f^{-1}, the given equation must be solved for x in terms of y. Let $x = f^{-1}(y)$.

Solving the given equation for x:

$$y = 3x + 2, \text{ where } y = f(x).$$

Subtract 2 from both sides of this equation:

$$y - 2 = 3x + 2 - 2$$
$$y - 2 = 3x.$$

Divide both sides of this equation by 3:

$$\frac{y-2}{3} = \frac{3x}{3}$$
$$\frac{y-2}{3} = x$$

or $\qquad x = \dfrac{y}{3} - \dfrac{2}{3}$

Hence, the inverse function f^{-1} given by:

$$x = f^{-1}(y) = \frac{y}{3} - \frac{2}{3}$$

$$\text{or} \quad x = f^{-1}(y) = \frac{1}{3}y - \frac{2}{3}.$$

Of course, the letter that we use to denote a number in the domain of the inverse function is of no importance whatsoever, so this last equation can be rewritten $f^{-1}(u) = \dfrac{1}{3}u - \dfrac{2}{3},$ or $f^{-1}(s) = \dfrac{1}{3}s - \dfrac{2}{3},$ and it will still define the same function f^{-1}. Note that we could have interchanged x and y in the original function as we did in the previous example.

 Given the function f defined by the equation

$$y = f(x) = \frac{3x + 4}{5},$$

where the domain (and the range) of f is the set R of all real numbers.

(a) Find the equation $x = g(y) = f^{-1}(x)$ that defines f^{-1}.

(b) Show that $f^{-1}(f(x)) = x$.

(c) Show that $f\left(f^{-1}(y)\right) = y$.

A (a) The definition of a function f is a set of ordered pairs (x,y) where

(1) x is an element of a set X

(2) y is an element of a set Y, and

(3) no two pairs in f have the same first element.

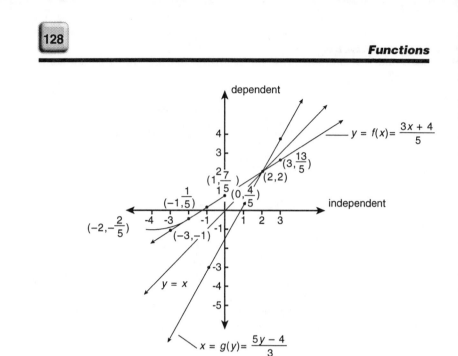

By definition, f is the infinite set of ordered pairs

$$\left\{ \left(x, \frac{3x+4}{5} \right) \Big| x \in R \right\},$$

which includes $(0, \frac{4}{5})$, $(2,2)$, $(7,5)$, $(12,8)$, $(-3, -1)$, etc. Furthermore, no two ordered pairs have the same first element. That is, for each element of X a unique value of Y is assigned. For example, if $x = 0$, we obtain only one y value, $\frac{4}{5}$.

We construct the following table to calculate the x and corresponding y values. Note that x is the independent variable and y is the dependent variable.

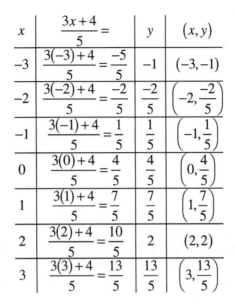

x	$\dfrac{3x+4}{5} =$	y	(x, y)
-3	$\dfrac{3(-3)+4}{5} = \dfrac{-5}{5}$	-1	$(-3, -1)$
-2	$\dfrac{3(-2)+4}{5} = \dfrac{-2}{5}$	$\dfrac{-2}{5}$	$\left(-2, \dfrac{-2}{5}\right)$
-1	$\dfrac{3(-1)+4}{5} = \dfrac{1}{5}$	$\dfrac{1}{5}$	$\left(-1, \dfrac{1}{5}\right)$
0	$\dfrac{3(0)+4}{5} = \dfrac{4}{5}$	$\dfrac{4}{5}$	$\left(0, \dfrac{4}{5}\right)$
1	$\dfrac{3(1)+4}{5} = \dfrac{7}{5}$	$\dfrac{7}{5}$	$\left(1, \dfrac{7}{5}\right)$
2	$\dfrac{3(2)+4}{5} = \dfrac{10}{5}$	2	$(2, 2)$
3	$\dfrac{3(3)+4}{5} = \dfrac{13}{5}$	$\dfrac{13}{5}$	$\left(3, \dfrac{13}{5}\right)$

See the accompanying figure, which shows the graph of the function f (which is also the graph of the equation $y = \dfrac{3x+4}{5}$). We can say that f carries (or maps) any real number x into the number

$$f : x \rightarrow \frac{3x+4}{5}.$$

Now to find the inverse function, we must find a function which takes each element of the original set Y and relates it to a unique value of X. There cannot be two values of X for a given value of Y in order for the inverse function to exist. That is, if this is true: (x_1, y) and (x_2, y), then there is no f^{-1}.

To find $x = g(y)$, we solve for x in terms of y.

Given: $y = \dfrac{3x+4}{5}$

Multiply both sides by 5,

$$5y = 3x + 4$$

Subtract 4 from both sides,

$$5y - 4 = 3x$$

Divide by 3 and solve for x,

$$x = \frac{5y - 4}{3} = f^{-1}(y) = g(y).$$

Choose y-values and find their corresponding x-values, as shown in the following table. Note that y is the independent variable and x is the dependent variable.

y	$g(y) = \dfrac{5y-4}{3} =$	x	(y,x)
-3	$\dfrac{5(-3)-4}{3}$	$\dfrac{-19}{3} = -6\dfrac{1}{3}$	$\left(-3, -6\dfrac{1}{3}\right)$
-2	$\dfrac{5(-2)-4}{3}$	$\dfrac{-14}{3} = -4\dfrac{2}{3}$	$\left(-2, -4\dfrac{2}{3}\right)$
-1	$\dfrac{5(-1)-4}{3}$	$\dfrac{-9}{3} = -3$	$(-1,-3)$
0	$\dfrac{5(0)-4}{3}$	$\dfrac{-4}{3} = -1\dfrac{1}{3}$	$\left(0, -1\dfrac{1}{3}\right)$
1	$\dfrac{5(1)-4}{3}$	$\dfrac{1}{3}$	$\left(1, \dfrac{1}{3}\right)$
2	$\dfrac{5(2)-4}{3}$	2	$(2,2)$
3	$\dfrac{5(3)-4}{3}$	$\dfrac{11}{3} = 3\dfrac{2}{3}$	$\left(3, 3\dfrac{2}{3}\right)$

See graph. Since there is only one value of x for each value of y, this equation defines the inverse function f^{-1}. The graph of f^{-1} is the image of the graph of f in the mirror $y = x$.

(b) Given $f(x) = \dfrac{3x + 4}{5} = y.$

Then perform the operation of f^{-1} on $y = f(x)$ where

$f^{-1}(x) = \dfrac{5y-4}{3}$. That is, substitute for $y : \dfrac{3x+4}{5}$.

$$f^{-1}(f(x)) = f^{-1}\left(\dfrac{3x+4}{5}\right) = \dfrac{5\left(\dfrac{3x+4}{5}\right) - 4}{3} = \dfrac{\dfrac{15x+20}{5} - 4}{3}$$

$$= \dfrac{\dfrac{15x+20-20}{5}}{3} = \dfrac{3x}{3} = x, \qquad \text{or}$$

$$f^{-1}(f(x)) = \dfrac{5f(x)-4}{3} = \dfrac{5\left(\dfrac{3x+4}{5}\right) - 4}{3} = x.$$

(c)   We now perform the operation of f on

$$f^{-1}(y) : \dfrac{5y-4}{3} = x. \quad \text{Note} \quad f(x) = \dfrac{3x+4}{5}$$

$$f(f^{-1}(y)) = f\left(\dfrac{5y-4}{3}\right) = \dfrac{3\left(\dfrac{5y-4}{3}\right) + 4}{5}$$

$$= \dfrac{5y-4+4}{5} = \dfrac{5y}{5} = y = f(x)$$

Since $f = \left\{\left(x, \dfrac{3x+4}{5}\right)\right\}$, this function f may be thought of as a sequence of directions listing the operations that must be performed on x to get $\dfrac{3x+4}{5}$. These operations are, in order: take any number x, multiply it by 3, add 4, and then divide by 5. The inverse function

$$f^{-1} = \left\{\left(y, \dfrac{5y-4}{3}\right)\right\}$$

tells us to multiply by 5, subtract 4, and then divide by 3. This "undoes," in reverse order, the operations performed by f. The function f^{-1} could be called "the undoing function" because it undoes what the function f has done.

5.4 Increasing and Decreasing Functions

On intervals where the graph moves upward from left to right, the function is said to be increasing. On intervals where the graph moves downward from left to right, the function is said to be decreasing. The turning points are those points where the function changes from increasing to decreasing or from decreasing to increasing. There are two kinds of turning points: relative maxima (high points) and relative minima (low points). Examine the graph below.

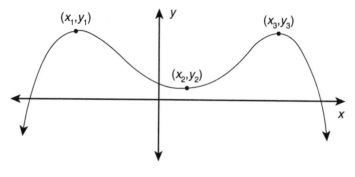

Figure 5.4

Using the vertical line test, it is easy to see that this is a graph of a function. It is an increasing function on the interval where $x \le x_1$ and on the interval where $x_2 < x < x_3$, and it is a decreasing function on the interval where $x_1 < x < x_2$ and where $x > x_3$. The relative maximum points are at (x_1, y_1) and (x_3, y_3), while the only relative minimum point is at (x_2, y_2).

Problem Solving Example:

 Given the function below:

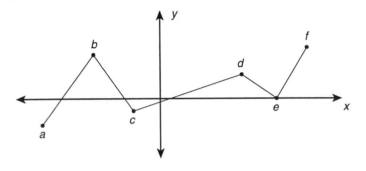

Over what intervals is the function increasing? Decreasing?

A The function increases when the line moves upward from left to right. Therefore the function increases at the intervals from a to b, from c to d, and from e to f.

In contrast, the function decreases when the line moves downward from left to right, or over the intervals from b to c and from d to e.

5.5 Polynomial Functions and Their Graphs

A polynomial in x is an expression of the form

$$a_n x^n + a_{n-1} x^{n-1} + \ldots + a_1 x + a_0,$$

where $a_1, a_2, \ldots$ and a_n are real numbers and where all the exponents are positive integers. When $a_n \neq 0$, this polynomial is said to be of degree n. It is common to let $P(x)$ represent

$$a_n x^n + a_{n-1} x^{n-1} + \ldots + a_1 x + a_0.$$

Then $y = P(x)$ is a polynomial function. A function with the property that

$$P(-x) = P(x)$$

is an even function, while a function with the property

$$P(-x) = -P(x)$$

is an odd function. Even functions are symmetric with respect to the y-axis, while odd functions are symmetric with respect to the origin.

It would be possible to obtain the graph of a polynomial function $y = P(x)$ by simply setting up a table and plotting a large number of points; this is how a computer or a graphing calculator operates. However, it is often desirable to have some basic information about the graph prior to plotting points. The graph of the polynomial function, $y = a_0$ is a line which is parallel to the x-axis and $|a_0|$ units above or below the x-axis, depending on whether a_0 is positive or negative. A function of this type is called a constant function. The graph of the polynomial function

$$y = a_1 x + a_0$$

is a line with slope a_1 and with a_0 as the y-intercept. The graph of the polynomial function

$$y = a_2 x^2 + a_1 x + a_0$$

is a parabola.

It is much more difficult to graph a polynomial function with degree greater than two. However, here are three items which should be investigated.

(1) Find lines (x-axis and y-axis) of symmetry and find out whether the origin is a point of symmetry. (See Section 2.4.)

(2) Find out about intercepts. The y-intercept is easy to find but the x-intercepts are usually much more difficult to identify. If possible, factor $P(x)$.

(3)   Find out what happens to $P(x)$ when $|x|$ is large.

This procedure is illustrated in the following example.

EXAMPLE

Graph

$$y = x^4 - 5x^2 + 4$$

(1)   The graph has symmetry with respect to the y-axis.

(2)   The y-intercept is at 4. Since

$$x^4 - 5x^2 + 4 = (x^2 - 4)(x^2 - 1)$$
$$= (x - 2)(x + 2)(x - 1)(x + 1),$$

the x-intercepts are at 1, 2, -1, and -2.

(3)   As $|x|$ gets large, $P(x)$ gets large.

Here is a sketch of the graph.

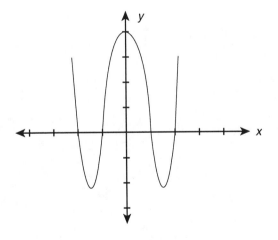

Figure 5.5

Problem Solving Example:

Graph the function
$$y = x^3 - 9x.$$

Choosing values of x in the interval $-4 \leq x \leq 4$, we have for $y = x^3 - 9x$,

x	-4	-3	-2	-1	0	1	2	3	4
y	-28	0	10	8	0	-8	-10	0	28

Notice that for each ordered pair (x,y) listed in the table there exists a pair $(-x,-y)$ which also satisfies the equation, indicating symmetry with respect to the origin. To prove that this is true for all

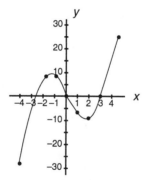

points on the curve, we substitute $(-x,-y)$ for (x,y) in the given equation and show that the equation is unchanged. Thus

$$-y = (-x)^3 - 9(-x) = -x^3 + 9x$$

or, multiplying each member by -1,

$$y = x^3 - 9x$$

which is the original equation.

The curve is illustrated in the figure. The domain and range of the function have no restrictions in the set of real numbers. The x-intercepts are found from

$$y = 0 = x^3 - 9x$$
$$0 = x(x^2 - 9)$$
$$0 = x(x-3)(x+3)$$
$$x = 0 \quad x - 3 = 0 \quad x + 3 = 0$$
$$x = 3 \qquad x = -3$$

The curve has three x-intercepts at $x = -3$, $x = 0$, $x = 3$. This agrees with the fact that a cubic equation can have three roots. The curve has a single y-intercept at $y = 0$ since for $x = 0$, $y = 0^3 - 9(0) = 0$.

5.6 Rational Functions and Their Graphs

When $P(x)$ and $Q(x)$ are polynomials,

$$y = \frac{P(x)}{Q(x)}$$

is called a rational function. The domain of this function is the set of all real numbers x with the property that $Q(x) \neq 0$.

Graphing rational functions is rather difficult. As is the case for polynomial functions, it is desirable to have a general procedure for graphing rational functions. Here is the suggested method for

$$y = \frac{P(x)}{Q(x)}$$

where $P(x) = a_n x^n + a_{n-1} x^{n-1} + \ldots + a_1 x + a_0$ and

$$Q(x) = b_m x^m + b_{m-1} x^{m-1} + \ldots + b_1 x + b_0.$$

(1) Find lines (x-axis and y-axis) of symmetry and determine whether the origin is a point of symmetry. (See Section 2.4.)

(2) Find out about intercepts. The y-intercept is at $\dfrac{a_0}{b_0}$ and the x-intercepts will be at values of x where $P(x) = 0$.

(3) Find vertical asymptotes. A line $x = c$ is a vertical asymptote whenever $Q(c) = 0$, and $P(c) \neq 0$.

(4) Find horizontal asymptotes

    (a) If $m = n$, then $y = \dfrac{a_n}{b_m}$ is the horizontal asymptote.

    (b) If $m > n$, then $y = 0$ is the horizontal asymptote.

    (c) If $m < n$, then there is no horizontal asymptote.

This procedure is illustrated in the following example.

EXAMPLE

Graph

$$y = \frac{x}{(x-1)(x+3)}$$

(1) The axes are not lines of symmetry, nor is the origin a point of symmetry.

(2) The x-intercept and the y-intercept are both at the origin.

(3) The lines $x = 1$ and $x = -3$ are both vertical asymptotes.

(4) The line $y = 0$ is the horizontal asymptote.

Here is a sketch of the graph.

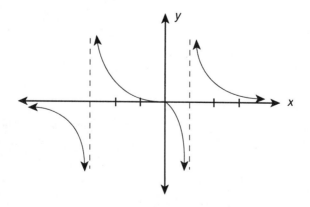

Figure 5.6

Problem Solving Example:

 Graph $y = \dfrac{(x+1)}{(x+2)}$

A 1. The axes are not lines of symmetry, nor is the origin a point of symmetry.

2. The x-intercept is at -1 or $(-1,0)$, while the y-intercept is at $\dfrac{1}{2}$ or $(0, \dfrac{1}{2})$.

3. The line $x = -2$ is a vertical asymptote, while the horizontal asymptote is located at the line $y = 1$.

4. A sketch of the graph is:

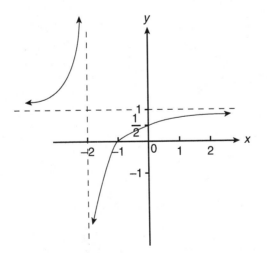

5.7 Special Functions and Their Graphs

It is possible to define a function by using different rules for different portions of the domain. The graphs of such functions are determined by graphing the different portions separately. Here is an example.

EXAMPLE

Graph

$$f(x) = \begin{cases} x & \text{if } x \le 1 \\ 2x & \text{if } x > 1 \end{cases}$$

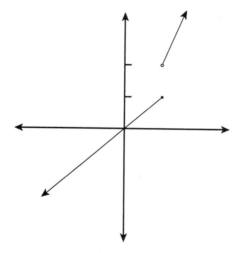

Figure 5.7

Notice that point (1,1) is part of the graph, but (1, 2) is not.

Functions which involve absolute value can often be completed by translating them to a two-rule form. Consider this example.

EXAMPLE

Graph

$$f(x) = |x| - 1$$

Since $|x| = \begin{cases} x & \text{if } x \geq 0 \\ -x & \text{if } x < 0 \end{cases}$

$f(x)$ can be translated to the following form.

$$f(x) = \begin{cases} x - 1 & \text{if } x \geq 0 \\ -x - 1 & \text{if } x < 0 \end{cases}$$

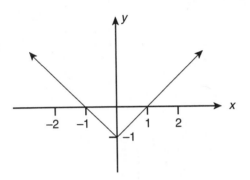

Figure 5.8

The greatest integer function, denoted by $f(x) = [|x|]$, is defined by $f(x) = j$, where j is the integer with the property that $j \leq x < j + 1$. The graph of this function follows.

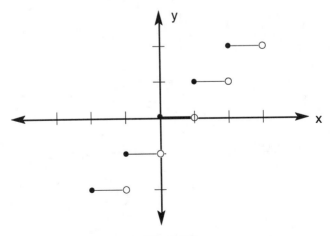

Figure 5.9

Problem Solving Examples:

 Sketch the graphs of the following functions:

(a) $f(x) = \begin{cases} x & \text{if } x \le 4 \\ \frac{1}{2}x - 2 & \text{if } x > 4 \end{cases}$

(b) $f(x) = |4x + 9|$

(c) $f(x) = \dfrac{|x|}{x+1}$ if $x \ne -1$

A (a)

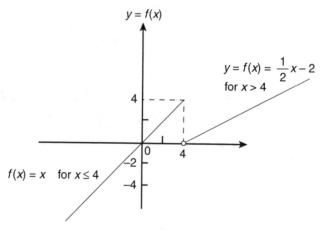

(b)

$$f(x) = |4x + 9| = \begin{cases} 4x + 9 & x > -\dfrac{9}{4} \\ 0 & x = -\dfrac{9}{4} \\ -(4x + 9) & x < -\dfrac{9}{4} \end{cases}$$

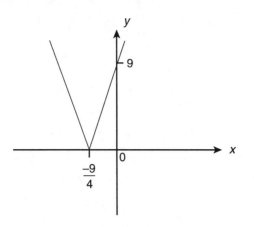

In this case we are graphing the equation of a straight line. However, because the equation is found under the absolute value signs, our line never crosses the *x*-axis. When our values for *y* would have been negative, the absolute value is taken — so we get a positive *y*.

(c)

$$f(x) = \frac{|x|}{x+1} = \begin{cases} \dfrac{x}{x+1} & x > 0 \\ 0 & x = 0 \\ -\dfrac{x}{x+1} & x < 0, \quad x \neq -1 \end{cases}$$

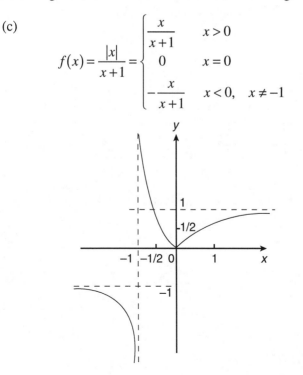

Draw the graph of the step function

$$y = \begin{cases} 2 & -2 \le x < -1 \\ 1 & -1 \le x < 0 \\ 0 & 0 \le x < 1 \\ 1 & 1 \le x < 2 \\ 2 & 2 \le x < 3 \end{cases}$$

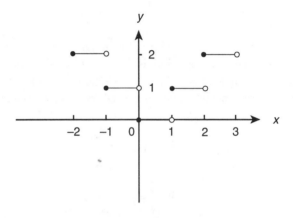

Quiz: Functions

1. Let $f(x) = \dfrac{2x+3}{x-5}$. Which of the following is true?

    (A) $f^{-1}(x) = \dfrac{5x+2}{x-3}$.

    (B) $f^{-1}(x) = \dfrac{2x-3}{x+5}$.

    (C) $f^{-1}(x) = \dfrac{2x-5}{x+3}$.

    (D) $f^{-1}(x) = \dfrac{3+2x}{5-x}$.

    (E) $f^{-1}(x) = \dfrac{5x+3}{x-2}$.

2. $g(x) = x^2 + 4$, find $g(-3)$

    (A) -3.   (D) 3.

    (B) 4.   (E) 13.

    (C) -5.

3. If $f(x) = 7x^2 + 3$ and $g(x) = 2x - 9$, $g(f(2)) =$

    (A) 28.   (D) 19.

    (B) 0.   (E) 53.

    (C) 31.

4. Of the following relations, the ones that are functions are

    I. $\dfrac{x^2}{81} - \dfrac{y^2}{16} = 3$

    II. $x^2 + \left| \dfrac{\sqrt{y^2}}{3} \right| = 3y$

    III. $y = \sqrt{3}x$

    (A) I.   (D) I, II, and III.

    (B) I and III.   (E) II and III.

    (C) II.

5. On what intervals(s) is $f(x) = x^3 + 2x^2$ increasing?

    (A) $\left(-\dfrac{4}{3}, \infty \right)$.   (D) $\left(-\dfrac{4}{3}, 0 \right)$.

    (B) $(0, \infty)$.   (E) $\left(-\infty, -\dfrac{4}{3} \right)$.

    (C) $\left(-\infty, -\dfrac{4}{3} \right)$ and $(0, \infty)$.

6. If a function is defined at $|2 - 5x| < 3$, then the interval which does not contain any solution for x is

   (A) $0 < x < 1$.

   (B) $0 < x < 2$.

   (C) $-\dfrac{1}{25} < x < 0$.

   (D) $x < -\dfrac{1}{5}, \ x > 1$.

   (E) $-1 < x < 1$.

7. Which of the following functions are neither odd nor even?

   (A) $f(x) = x^4 + x^2 + 1$.

   (B) $f(x) = x^5 + x^3 + x + 1$.

   (C) $f(x) = (x + 2)^2$.

   (D) $f(x) = x^2 + 4$.

   (E) $f(x) = x^3 + x$.

8. What is the domain of $f(x) = \dfrac{2x + 1}{x^3 - x}$?

   (A) all real numbers except $0, 1, -1$.

   (B) all real numbers except $0, 1$.

   (C) all real numbers except $1, -1$.

   (D) all real numbers except 0.

   (E) all real numbers.

9. Let $f(x)$ be defined as follows:

$$f(x) = \begin{cases} x & x < -5 \\ 0 & -5 \leq x < 0 \\ [[x]] & x \geq 0 \end{cases}$$

Find $f(1.1)$

   (A) 0.   (D) 1.

   (B) 1.1.   (E) 2.

   (C) –5.

10. The figure is a sketch of $y = f(x)$. What is $f(f(6))$?

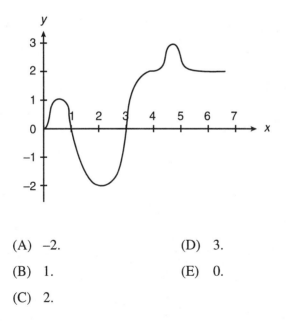

   (A) –2.   (D) 3.

   (B) 1.   (E) 0.

   (C) 2.

ANSWER KEY

1.	(E)	6.	(D)
2.	(E)	7.	(B)
3.	(E)	8.	(A)
4.	(E)	9.	(D)
5.	(C)	10.	(A)

CHAPTER 6

Trigonometry

6.1 Trigonometric Functions

An angle is said to be in standard position when its initial side corresponds to the positive x-axis. In the picture below, α is in standard position, and (x,y) denotes a point on the terminal side of α.

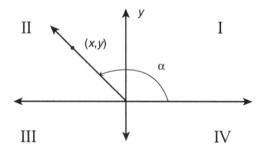

Figure 6.1

Also, r is used to denote the distance between the origin and the point corresponding to (x,y), and

$$r = \sqrt{x^2 + y^2}.$$

The six trigonometric functions are named sine, cosine, tangent, cotangent, secant, and cosecant and are denoted by sin, cos, tan, cot, sec, and csc. These functions are defined as follows:

$$\sin\alpha = \frac{y}{r} \qquad\qquad \cot\alpha = \frac{x}{y}, y \neq 0$$

$$\cos\alpha = \frac{x}{r} \qquad\qquad \sec\alpha = \frac{r}{x}, x \neq 0$$

$$\tan\alpha = \frac{y}{x}, x \neq 0 \qquad \csc\alpha = \frac{r}{y}, y \neq 0$$

From properties of similar triangles, it is easy to show that it does not make any difference which point is selected on the terminal side of α. An angle is said to be in a particular quadrant when the terminal side of the angle falls in that quadrant. Quadrants are shown in figure 6.1.

Here is a table indicating whether the trigonometric functions are positive (+) or negative (−) in particular quadrants.

Quadrant for α	x	y	sin α	cos α	tan α	cot α	sec α	csc α
I	+	+	+	+	+	+	+	+
II	−	+	+	−	−	−	−	+
III	−	−	−	−	+	+	−	−
IV	+	−	−	+	−	−	+	−

Table 6.1

From the definition of the functions, it is obvious that

$$\sec\alpha = \frac{1}{\cos\alpha}, \quad \csc\alpha = \frac{1}{\sin\alpha}, \quad \text{and } \cot\alpha = \frac{1}{\tan\alpha}$$

This is valuable information when calculators are used in the context of trigonometry. Most scientific calculators have only keys involving sin, cos, and tan functions. However, functional values of the secant, cosecant, and cotangent function can be found with a calculator by finding the corresponding functional values of reciprocals and using the $^1/_x$ key.

When calculators are used to find functional values of trigonometric functions, only approximations appear on the display of the calculator. However, there are some angles for which geometric properties can be used to find the exact functional values. A few of these are summarized in the table below.

α	$\sin \alpha$	$\cos \alpha$	$\tan \alpha$	$\cot \alpha$	$\sec \alpha$	$\csc \alpha$
$0°$	0	1	0	undefined	1	undefined
$30°$	$\dfrac{1}{2}$	$\dfrac{\sqrt{3}}{2}$	$\dfrac{\sqrt{3}}{3}$	$\sqrt{3}$	$\dfrac{2\sqrt{3}}{3}$	2
$45°$	$\dfrac{\sqrt{2}}{2}$	$\dfrac{\sqrt{2}}{2}$	1	1	$\sqrt{2}$	$\sqrt{2}$
$60°$	$\dfrac{\sqrt{3}}{2}$	$\dfrac{1}{2}$	$\sqrt{3}$	$\dfrac{\sqrt{3}}{3}$	2	$\dfrac{2\sqrt{3}}{3}$
$90°$	1	0	undefined	0	undefined	1
$180°$	0	-1	0	undefined	-1	undefined
$270°$	-1	0	undefined	0	undefined	-1

Table 6.2

This table is given with angles measured in degrees. Since $360°$ = 2π radians, it is easy to translate the table into corresponding radian measure categories.

The trigonometric functions are periodic. The graph of each function is shown on the next page.

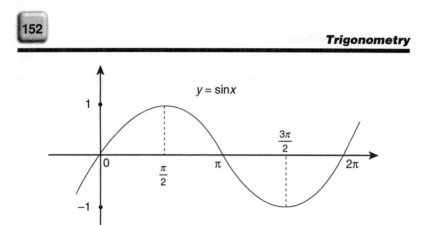

Figure 6.2

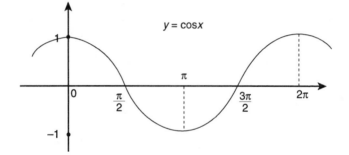

Figure 6.3

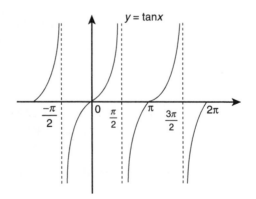

Figure 6.4

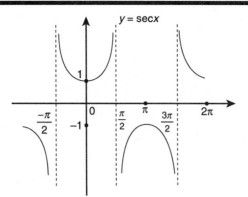

Figure 6.5

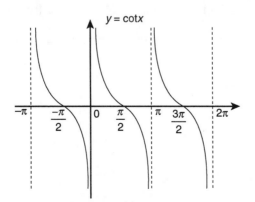

Figure 6.6

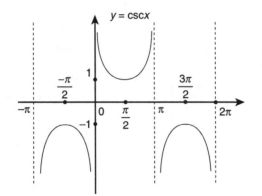

Figure 6.7

Graphs of equations of the form

$$y = \sin(x + a)$$

are also important. To illustrate the effect of replacing "x" by "$x + a$," the graphs of

$$y = \sin x \quad \text{and} \quad y = \sin\left(x + \frac{\pi}{2}\right)$$

are given. Notice that the graph of $y = \sin\left(x + \frac{\pi}{2}\right)$ is a "copy" of $y = \sin x$ which has been moved $\frac{\pi}{2}$ units to the left.

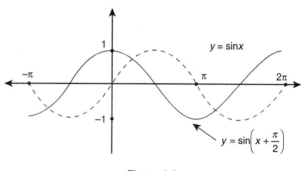

Figure 6.8

The graphs of $y = \sin x$ and $y = 2 \sin x$ are pictured below. Notice that the amplitude for $y = \sin x$ is 1 while the amplitude for $y = 2 \sin x$ is 2. (The amplitude is the maximum height of the curve above the x-axis. A more general definition is amplitude $y = \frac{1}{2}(y_{max} - y_{min})$.)

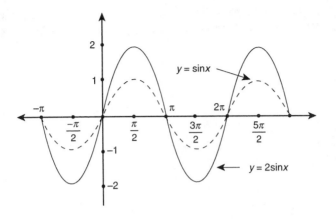

Figure 6.9

The graphs of $y = \sin x$ and $y = \sin 2x$ are pictured below. Notice that $y = \sin x$ is periodic with a period of 2π and $y = \sin 2x$ is periodic with a period of π.

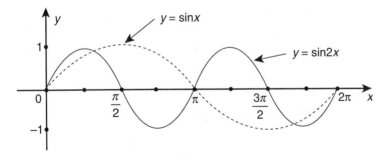

Figure 6.10

Problem Solving Examples:

 Complete the following table:

	1	2	3	4	5	6	7	8	9
Measure of θ in radians	0	$\frac{1}{6}\pi$	$\frac{1}{4}\pi$		$\frac{1}{2}\pi$	$\frac{2}{3}\pi$			π
Measure of θ in degrees	0°			60°			135°	150°	

If an angle θ is A degrees and also t radians, then the numbers A and t are related by the equation:

$$\frac{A}{180°} = \frac{t}{\pi} \qquad (1)$$

Thus, equation (1) can be used to complete the table. For column (2):

$$\frac{A}{180°} = \frac{\frac{1}{6}\pi}{\pi}$$

$$\frac{A}{180°} = \frac{1}{6}$$

Multiplying both sides by 180°,

$$180°\left(\frac{A}{180°}\right) = 180°\left(\frac{1}{6}\right)$$

$$A = 30°$$

For column (3):

$$\frac{A}{180°} = \frac{\frac{1}{4}\pi}{\pi}$$

$$\frac{A}{180°} = \frac{1}{4}$$

Multiplying both sides by 180°,

$$180°\left(\frac{A}{180°}\right) = 180°\left(\frac{1}{4}\right)$$
$$A = 45°$$

For column (4):

$$\frac{60°}{180°} = \frac{t}{\pi}$$
$$\frac{1}{3} = \frac{t}{\pi}$$

Multiplying both sides by π,

$$\pi\left(\frac{1}{3}\right) = \pi\left(\frac{1}{\pi}\right)$$
$$\frac{1}{3}\pi = t$$

For column (5):

$$\frac{A}{180°} = \frac{\frac{1}{2}\pi}{\pi}$$
$$\frac{A}{180°} = \frac{1}{2}$$

Multiplying both sides by 180°,

$$180°\left(\frac{A}{180°}\right) = 180°\left(\frac{1}{2}\right)$$

$$A = 90°$$

For column (6):

$$\frac{A}{180°} = \frac{\frac{2}{3}\pi}{\pi}$$

$$\frac{A}{180°} = \frac{2}{3}$$

Multiplying both sides by 180°,

$$180°\left(\frac{A}{180°}\right) = 180°\left(\frac{2}{3}\right)$$

$$A = 120°$$

For column (7):

$$\frac{135°}{180°} = \frac{t}{\pi}$$

$$\frac{27}{36} = \frac{t}{\pi}$$

$$\frac{3}{4} = \frac{t}{\pi}$$

Multiplying both sides by π,

$$\pi\left(\frac{3}{4}\right) = \pi\left(\frac{t}{\pi}\right)$$

$$\frac{3}{4}\pi = t$$

For column (8):

$$\frac{150°}{180°} = \frac{t}{\pi}$$

$$\frac{50}{60} = \frac{t}{\pi}$$

$$\frac{5}{6} = \frac{t}{\pi}$$

Multiplying both sides by π,

$$\pi\left(\frac{5}{6}\right) = \pi\left(\frac{t}{\pi}\right).$$

$$\frac{5}{6}\pi = t$$

For column (9):

$$\frac{A}{180°} = \frac{\pi}{\pi}$$

$$\frac{A}{180°} = 1$$

Multiplying both sides by $180°$,

$$180°\left(\frac{A}{180°}\right) = 180°(1)$$

$$A = 180°$$

All of the computed values are now put into the table as follows:

	1	2	3	4	5	6	7	8	9
Measure of θ in radians	0	$\frac{1}{6}\pi$	$\frac{1}{4}\pi$	$\frac{1}{3}\pi$	$\frac{1}{2}\pi$	$\frac{2}{3}\pi$	$\frac{3}{4}\pi$	$\frac{5}{6}\pi$	π
Measure of θ in degrees	0°	30°	45°	60°	90°	120°	135°	150°	180°

 Graph (1) $y = \sin x$

(2) $y = 4 \sin x$

(3) $y = \sin 4x$

(4) $y = \sin \left(x + \dfrac{\pi}{4} \right)$

(5) $y = A \sin(Bx + C) + D$

where A, B, C, D are real constants.

A (1)

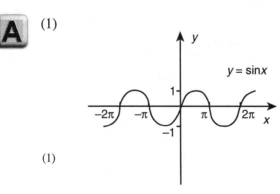

(1)

$y = \sin x$ is a periodic function with period $T = 2\pi$. Its amplitude is:

$$\frac{1}{2}\left[(y_{max}) - (y_{min})\right] = \frac{1}{2}\left[1 - (-1)\right] = 1.$$

The graph is symmetrical with respect to the origin.

(2) The graph of $y = 4 \sin x$ is the same as $y = \sin x$ except that the amplitude is

$$\frac{1}{2}\left[(y_{max}) - (y_{min})\right] = \frac{1}{2}\left[4 - (-4)\right] = 4.$$

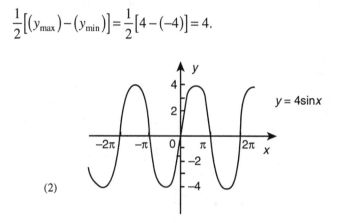

(2)

(3) $y = \sin 4x$ has an amplitude 1, but a period of

$$T = \frac{2\pi}{4} = \frac{\pi}{2}.$$

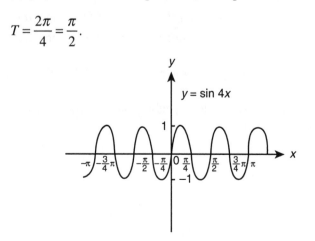

(4) $y = \sin\left(x + \dfrac{\pi}{4}\right)$ has period $T = 2\pi$, amplitude 1, and phase

shift of $\dfrac{\pi}{4}$ $\left(\dfrac{\pi}{4}$ is also called the phase angle$\right)$.

Note that the graph of $y = \sin\left(x + \dfrac{\pi}{4}\right)$ can be obtained by simply

shifting the graph of $y = \sin x$ by $\dfrac{\pi}{4}$ to the left of origin.

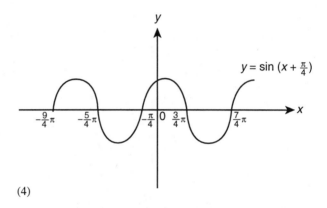

(4)

(5) For $y = A \sin(Bx + C) + D$, the constant $|A|$ is the amplitude of the function, the constant B decides the period of the function, $\left(\dfrac{2\pi \text{ radians}}{|B|}\right)$, $\dfrac{C}{B}$ is the phase angle, and D will shift the graph of $y = A \sin(Bx + C)$ up (or down) along the y-axis by D units for positive (or negative) D.

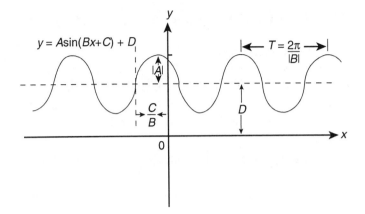

$y = A\sin(Bx+C) + D$

$T = \dfrac{2\pi}{|B|}$

$|A|$

$\dfrac{C}{B}$

D

Q Sketch three periods of the graph $y = 3\cos 2x$.

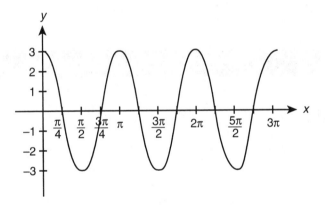

A The coefficient of the function is 3, which means that the maximum and minimum values are 3 and -3, respectively. The period of the cosine function is 2π radians divided by the coefficient of x. Therefore, the period of the cosine function given in this problem is

$$\left(\frac{2\pi \text{ radians}}{2}\right) = \pi \text{ radians}$$

and with this knowledge, we sketch the curve as in the figure.

6.2 Inverse Trigonometric Functions

A function has an inverse if and only if it is one-to-one. Since the trigonometric functions are periodic and, thus, not one-to-one, inverses can be established for trigonometric functions only when the domains are restricted. Here are the definitions of the inverse functions for the sine, cosine, and tangent functions.

$$y = \text{Arcsin } x \text{ if and only if } \sin y = x \text{ and } -\frac{\pi}{2} \leq y \leq \frac{\pi}{2}$$

$$y = \text{Arccos } x \text{ if and only if } \cos y = x \text{ and } 0 \leq y \leq \pi$$

$$y = \text{Arctan } x \text{ if and only if } \tan y = x \text{ and } -\frac{\pi}{2} < x < \frac{\pi}{2}$$

Arctan x is sometimes denoted $\tan^{-1}x$. Similarly $\sin^{-1}x$ and $\cos^{-1}x$ are defined.

Problem Solving Examples:

In $\triangle ABC$, $A = \arccos\left(-\frac{\sqrt{3}}{2}\right)$. What is the value of A expressed in radians?

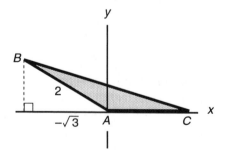

The expression "$\arccos\left(-\frac{\sqrt{3}}{2}\right)$" means "the angle whose cosine equals $-\dfrac{\sqrt{3}}{2}$." Angles whose cosine equals $-\dfrac{\sqrt{3}}{2}$ are 150°, 210°, −150°, and −210°.

Since the principal value of an arccosine of an angle is the positive angle having the smallest numerical value of the angle, 150°, or $\dfrac{5\pi}{6}$, is the principal value of angle A.

Q Evaluate: (a) $\sin^{-1}\dfrac{\sqrt{3}}{2}$, (b) $\tan^{-1}\left(-\sqrt{3}\right)$.

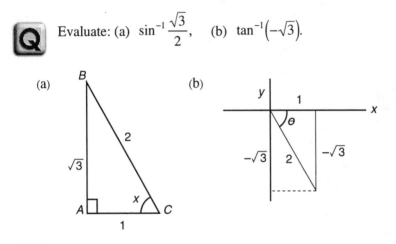

A (a) Recall that inverse sines are angles. Thus we are looking for the angle whose sin is $\dfrac{\sqrt{3}}{2}$. $\sin^{-1}\dfrac{\sqrt{3}}{2} = x$ means $\sin x = \dfrac{\sqrt{3}}{2}$ where $\sin = \dfrac{\text{opposite}}{\text{hypotenuse}}$.

We note that triangle ABC is a 30°–60° right triangle, and angle $x = 60°$. Since $\sin 60° = \dfrac{\sqrt{3}}{2}$,

$$\sin^{-1} = \frac{\sqrt{3}}{2} = 60°.$$

(b) Recall that inverse tangents are angles. Thus we are looking for the angle whose tangent is $-\sqrt{3}$. $\mathrm{Tan}^{-1}\left(-\sqrt{3}\right) = \theta$ means

$\tan \theta = -\sqrt{3}$ where $\tan = \dfrac{\text{opposite}}{\text{adjacent}}$.

Since tangent is negative in the 4th quadrant, we draw our triangle there, and note it is a 30°–60° right triangle, and angle $\theta = (-60°)$. Since $\tan(-60°) = \dfrac{-\sqrt{3}}{1}$, $\tan^{-1}\left(-\sqrt{3}\right) = -60°$.

Q Evaluate cos[arcsin (–1)].

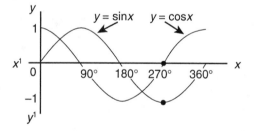

A The expression "arcsin (–1)" means "the angle whose sine equals –1." Between 0° and 360°, the only angle whose sine equals –1 is 270°.

Hence, cos[arcsin (–1)] = cos 270°.

The value of cos 270° = 0.

Note: In problems of this type, a sketch of $y = \sin x$ and $y = \cos x$ is very useful.

Q Evaluate $\tan\left[\dfrac{1}{2}\arcsin\left(-\dfrac{8}{17}\right)\right]$.

A   Let $\theta = \arcsin\left(-\dfrac{8}{17}\right)$. Then $\sin\theta = -\dfrac{8}{17}$. Thus, we wish to

evaluate $\tan\left(\dfrac{\theta}{2}\right)$. Using the half-angle formula for tan, (which can be derived from trigonometric identities listed in the next section) we obtain:

$$\tan\left(\frac{\theta}{2}\right) = \frac{1-\cos\theta}{\sin\theta}.$$

Thus, we must find, in addition to $\sin\theta = -\dfrac{8}{17}$, the corresponding value of cos θ. Observe that when $\sin\theta$ is negative, θ must be a negative angle in the fourth quadrant, since we have the restriction on the inverse sin function $-\dfrac{\pi}{2} \le \theta \le \dfrac{\pi}{2}$. Thus, cos θ will be positive (since cos is positive in the fourth quadrant), and using the identity $\sin^2\theta + \cos^2\theta = 1$ we obtain:

$$\cos\theta = \sqrt{1-\sin^2\theta} = \sqrt{1-\left(-\frac{8}{17}\right)^2} = \sqrt{1-\frac{64}{289}}$$

$$= \sqrt{\frac{225}{289}} = \frac{15}{17}.$$

Therefore the desired value is

$$\tan\left(\frac{\theta}{2}\right) = \frac{1-\dfrac{15}{17}}{-\dfrac{8}{17}} = \frac{2}{17} \times \left(-\frac{17}{8}\right) = -\frac{1}{4},$$

or

$$\tan\left[\frac{1}{2}\arcsin\left(-\frac{8}{17}\right)\right] = -\frac{1}{4}.$$

6.3 Trigonometric Identities

Here is a list of trigonometric identities which are often useful:

$$\tan\alpha = \frac{\sin\alpha}{\cos\alpha}$$

$$\csc\alpha = \frac{1}{\sin\alpha}$$

$$\sec\alpha = \frac{1}{\cos\alpha}$$

$$\cot\alpha = \frac{1}{\tan\alpha}$$

$$\sin^2\alpha + \cos^2\alpha = 1$$

$$\tan^2\alpha + 1 = \sec^2\alpha$$

$$\cot^2\alpha + 1 = \csc^2\alpha$$

$$\sin(-\alpha) = -\sin\alpha$$

$$\cos(-\alpha) = \cos\alpha$$

$$\tan(-\alpha) = -\tan\alpha$$

$$\sin(\alpha \pm \beta) = \sin\alpha\cos\beta \pm \cos\alpha\sin\beta$$

$$\cos(\alpha \pm \beta) = \cos\alpha\cos\beta \pm \sin\alpha\sin\beta$$

$$\tan(\alpha \pm \beta) = \frac{\tan\alpha \pm \tan\beta}{1 \pm \tan\alpha\tan\beta}$$

$$\sin 2\alpha = 2\sin\alpha\cos\alpha$$

$$\cos 2\alpha = \cos^2\alpha - \sin^2\alpha$$

$$\tan 2\alpha = \frac{2\tan\alpha}{1 - \tan^2\alpha}$$

$$\sin^2\alpha = \frac{1 - \cos 2\alpha}{2}$$

$$\cos^2\alpha = \frac{1 + \cos 2\alpha}{2}$$

Problem Solving Examples:

Find the exact value for

(1) $\sin 75° - \sin 15°$

(2) $\sin^2 22.5°$

(3) $\tan^2 15°$

(1) By use of the formula for the sine of a sum, one obtains

$\sin 75° - \sin 15° = \sin(45° + 30°) - \sin(45° - 30°)$

$= \sin 45° \cos 30° + \cos 45° \sin 30° - (\sin 45° \cos 30° - \cos 45° \sin 30°)$

$= 2 \cos 45° \sin 30°$

$= 2\left(\dfrac{\sqrt{2}}{2}\right)\left(\dfrac{1}{2}\right)$

$= \dfrac{\sqrt{2}}{2}$

(2) The formula for the square of the sine of an angle in terms of a function of twice the angle is

$$\sin^2 \alpha = \frac{1 - \cos 2\alpha}{2} \quad \text{(see identities)}$$

For $\alpha = 22.5°, 2\alpha = 45°$.

$$\text{So, } \sin^2 22.5° = \frac{1 - \cos 45°}{2}$$

$$= \left[1 - \frac{\sqrt{2}}{2}\right] \times \frac{1}{2}$$

$$= \frac{1}{2} - \frac{\sqrt{2}}{4}$$

$$= \frac{2 - \sqrt{2}}{4}$$

(3) The formulas for the squares of the sine and cosine of an angle in terms of a function of twice the angle are

$$\sin^2 \alpha = \frac{1 - \cos 2\alpha}{2} \quad \text{(See identities.)}$$

$$\cos^2 \alpha = \frac{1 + \cos 2\alpha}{2}$$

Now, $\tan^2 \alpha \frac{\sin^2 \alpha}{\cos^2 \alpha} = \dfrac{\dfrac{1 - \cos 2\alpha}{2}}{\dfrac{1 + \cos 2\alpha}{2}} = \dfrac{1 - \cos 2\alpha}{1 + \cos 2\alpha}$

For $\alpha = 15°$, $2\alpha = 30°$.

So, $\tan^2 15° = \dfrac{1 - \cos 30°}{1 + \cos 30°}$

$$= \frac{1 - \dfrac{\sqrt{3}}{2}}{1 + \dfrac{\sqrt{3}}{2}} = \frac{2 - \sqrt{3}}{2 + \sqrt{3}}$$

$$= \left(2 - \sqrt{3}\right)^2 \quad \text{(Rationalizing the denominator.)}$$

 Find sin 15°, cos 15°, tan 15°, and cot 15°.

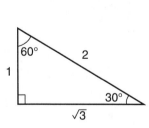

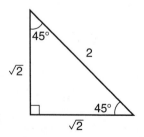

A To find the values of these trigonometric functions, use the subtraction formulas for the sine, cosine, tangent, and cotangent functions.

$$\sin(\alpha - \beta) = \sin \alpha \cos \beta - \cos \alpha \sin \beta$$

$$\cos(\alpha - \beta) = \cos \alpha \cos \beta + \sin \alpha \sin \beta$$

$$\tan(\alpha - \beta) = \frac{\tan \alpha - \tan \beta}{1 + \tan \alpha \tan \beta}$$

$$\cot(\alpha - \beta) = \frac{\cot \alpha \cot \beta + 1}{\cot \beta - \cot \alpha}$$

Recall that in a 30°–60° right triangle:

$$\sin 30° = \frac{1}{2} \text{ and } \cos 30° = \frac{\sqrt{3}}{2},$$

$$\sin 60° = \frac{\sqrt{3}}{2} \text{ and } \cos 60° = \frac{1}{2}.$$

In a 45°–45° right triangle:

$$\sin 45° = \frac{\sqrt{2}}{2} \text{ and } \cos 45° = \frac{\sqrt{2}}{2} \text{ (see figures)}$$

Now, substitute $\alpha = 45°$ and $\beta = 30°$ into the appropriate formulas.

$$\sin 15° = \sin(45° - 30°) = \sin 45° \cos 30° - \cos 45° \sin 30°$$

$$= \left(\frac{\sqrt{2}}{2}\right)\left(\frac{\sqrt{3}}{2}\right) - \left(\frac{\sqrt{2}}{2}\right)\left(\frac{1}{2}\right)$$

$$= \frac{\sqrt{2}\sqrt{3}}{4} - \frac{\sqrt{2}}{4} = \frac{\sqrt{2}(\sqrt{3} - 1)}{4}$$

$$\cos 15° = \cos(45° - 30°) = \cos 45° \cos 30° + \sin 45° \sin 30°$$

$$= \left(\frac{\sqrt{2}}{2}\right)\left(\frac{\sqrt{3}}{2}\right) + \left(\frac{\sqrt{2}}{2}\right)\left(\frac{1}{2}\right)$$

$$= \frac{\sqrt{2}\sqrt{3}}{4} + \frac{\sqrt{2}}{4} = \frac{\sqrt{2}\left(\sqrt{3}+1\right)}{4}$$

$$\tan 15° = \frac{\tan 45° - \tan 30°}{1 + \tan 45° \tan 30°}$$

$$= \frac{1 - \dfrac{\sqrt{3}}{3}}{1 + 1\left(\dfrac{\sqrt{3}}{3}\right)} = \frac{1 - \dfrac{\sqrt{3}}{3}}{1 + \dfrac{\sqrt{3}}{3}}$$

Obtaining a common denominator of 3 for both the numerator and the denominator,

$$\tan 15° = \frac{\dfrac{3}{3} - \dfrac{\sqrt{3}}{3}}{\dfrac{3}{3} + \dfrac{\sqrt{3}}{3}} = \frac{\dfrac{3 - \sqrt{3}}{3}}{\dfrac{3 + \sqrt{3}}{3}} = \frac{3 - \sqrt{3}}{3} \cdot \frac{3}{3 + \sqrt{3}} = \frac{3 - \sqrt{3}}{3 + \sqrt{3}}$$

$$= 2 - \sqrt{3} \quad \text{(Rationalizing the denominator.)}$$

$$\cot 15° = \cot\left(45° - 30°\right)$$

$$= \frac{\cot 45° \cot 30° + 1}{\cot 30° - \cot 45°}$$

$$= \frac{(1)\left(\sqrt{3}\right) + 1}{\sqrt{3} - 1}$$

$$= \frac{\sqrt{3} + 1}{\sqrt{3} - 1}$$

$$= 2 + \sqrt{3} \quad \text{(Rationalizing the denominator.)}$$

Change $\tan \theta (\sin \theta + \cot \theta \cos \theta)$ to $\sec \theta$.

 Distribute to obtain,

$\tan \theta(\sin \theta + \cot \theta \cos \theta) = \tan \theta \sin \theta + \tan \theta \cot \theta \cos \theta$

Recall that $\cot \theta = 1/\tan \theta$, and replace $\cot \theta$ by $1/\tan \theta$:

$= \tan \theta \sin \theta + \tan \theta (1/\tan \theta) \cos \theta$

$= \tan \theta \sin \theta + \cos \theta$

Since $\tan \theta = \sin \theta/\cos \theta$ we may replace $\tan \theta$ by $\sin \theta/\cos \theta$:

$$= \frac{\sin \theta}{\cos \theta} \sin \theta + \cos \theta$$

$$= \frac{\sin^2 \theta}{\cos \theta} + \cos \theta.$$

To combine terms, we convert $\cos \theta$ into a fraction whose denominator is $\cos \theta$, thus

$$= \frac{\sin^2 \theta}{\cos \theta} + \left(\frac{\cos \theta}{\cos \theta} \right) \cdot \cos \theta.$$ (Note that $\cos \theta/\cos \theta$ equals one, so the equation is unaltered)

$$= \frac{\sin^2 \theta}{\cos \theta} + \frac{\cos^2 \theta}{\cos \theta}$$

$$= \frac{\sin^2 \theta + \cos^2 \theta}{\cos \theta}.$$

Recall the identity $\sin^2 \theta + \cos^2 \theta = 1$; hence,

$$= \frac{1}{\cos \theta}$$

$$= \sec \theta.$$

Reduce the expression $\dfrac{\tan x - \cot x}{\tan x + \cot x}$ to one involving only $\sin x$.

Since, by definition, $\tan x = \dfrac{\sin x}{\cos x}$ and

$$\cot x = \frac{1}{\tan x} = \frac{1}{\sin x / \cos x} = \frac{\cos x}{\sin x},$$

$$\frac{\tan x - \cot x}{\tan x + \cot x} = \frac{\dfrac{\sin x}{\cos x} - \dfrac{\cos x}{\sin x}}{\dfrac{\sin x}{\cos x} + \dfrac{\cos x}{\sin x}}$$

$$= \frac{\dfrac{\sin x(\sin x)}{\sin x(\cos x)} - \dfrac{\cos x(\cos x)}{\cos x(\sin x)}}{\dfrac{\sin x(\sin x)}{\sin x(\cos x)} + \dfrac{\cos x(\cos x)}{\cos x(\sin x)}}$$

$$= \frac{\dfrac{\sin^2 x - \cos^2 x}{\sin x \cos x}}{\dfrac{\sin^2 x + \cos^2 x}{\sin x \cos x}}$$

$$= \frac{\sin^2 x - \cos^2 x}{\sin x \cos x} \times \frac{\sin x \cos x}{\sin^2 x + \cos^2 x}$$

$$= \frac{\sin^2 x - \cos^2 x}{\sin^2 x + \cos^2 x}$$

Since $\sin^2 x + \cos^2 x = 1$ or $\cos^2 x = 1 - \sin^2 x$,

$$\frac{\tan x - \cot x}{\tan x + \cot x} = \frac{\sin^2 x - \cos^2 x}{\sin^2 x + \cos^2 x} = \frac{\sin^2 x - \cos^2 x}{1}$$
$$= \sin^2 x - \cos^2 x$$
$$= \sin^2 x - \left(1 - \sin^2 x\right)$$
$$= \sin^2 x - 1 + \sin^2 x$$
$$= 2\sin^2 x - 1.$$

6.4 Triangle Trigonometry

Trigonometry can be applied rather easily to right triangles. Suppose that right triangle ABC is placed in a coordinate axis position, so that α is in the standard position. (See Figure 6.11.)

Then, from the definition of the sine, cosine, and tangent functions,

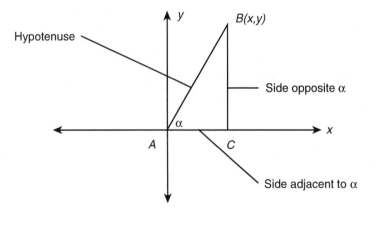

Figure 6.11

$$\sin \alpha = \frac{y}{r}$$

$$\cos \alpha = \frac{x}{r}$$

$$\tan \alpha = \frac{y}{x}$$

but x is the length of the side adjacent to α, y is the length of the side opposite α, and r is the length of the hypotenuse so

$$\sin \alpha = \frac{\text{length of side opposite } \alpha}{\text{length of hypotenuse}},$$

$$\cos \alpha = \frac{\text{length of side adjacent to } \alpha}{\text{length of hypotenuse}},$$

$$\tan \alpha = \frac{\text{length of side opposite } \alpha}{\text{length of side adjacent to } \alpha}.$$

These generalizations are useful when it is necessary to find measures of certain parts of right triangles given measures of other parts of the triangle (solving a triangle).

Here are two generalizations which can be applied in triangle-solving situations, when the triangle in question is not necessarily a right triangle.

If α, β, and γ are the measures of angles of a triangle and a, b, and c are the lengths of the sides opposite α, β, and γ respectively, then,

$$\frac{\sin \alpha}{a} = \frac{\sin \beta}{b} = \frac{\sin \gamma}{c} \quad \text{and}$$

$$c^2 = a^2 + b^2 - 2ab\cos\gamma.$$

The first of these generalizations is called the law of sines, and the second is called the law of cosines.

Determining the area of a triangle given (1) the length of two sides and the measure of the included angle, (2) the measures of two angles and the length of the included side, and (3) the length of three sides, can be accomplished by the use of the three formulas below.

If α, β, and γ are the measures of the angles of a triangle, and a, b, and c are the lengths of the sides opposite α, β, and γ respectively, and $s = \dfrac{1}{2}(a + b + c)$, then, the area ($K$) of the triangle is

$$K = \frac{1}{2}ab\sin\gamma$$

$$K = \frac{1}{2}a^2\frac{\sin\beta\sin\gamma}{\sin\alpha} \quad \text{and}$$

$$K = \sqrt{s(s-a)(s-b)(s-c)}$$

Problem Solving Examples:

 Find the values of the trigonometric functions of an angle of 300°.

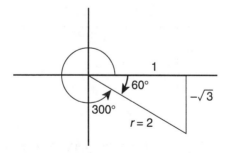

 An angle of 300° is a fourth quadrant angle and its reference angle is an angle of 60°. In the fourth quadrant the sine, tangent, cotangent and cosecant functions are negative. This yields:

$$\sin 300° = -\sin 60° = -\frac{\sqrt{3}}{2},$$

$$\cos 300° = \cos 60° = \frac{1}{2},$$

$$\tan 300° = -\tan 60° = -\sqrt{3},$$

$$\cot 300° = -\cot 60° = -\frac{\sqrt{3}}{3},$$

$$\sec 300° = \sec 60° = \frac{1}{\cos 60°} = \frac{1}{\frac{1}{2}} = 2, \quad \text{and}$$

$$\csc 300° = -\csc 60° = -\frac{1}{\sin 60°} = -\frac{1}{-\sqrt{3}/2} = -\frac{2}{\sqrt{3}}$$

$$= -\frac{2\sqrt{3}}{\sqrt{3}\sqrt{3}} = -\frac{2\sqrt{3}}{3}.$$

 Given that $\tan \theta = 2$ and $\cos \theta$ is negative, find the other functions of θ.

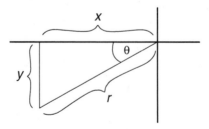

A Since $\cos \theta$ is negative, θ must be a second or third quadrant angle. In the second quadrant, the tangent function is negative. Hence, θ must be a third quadrant angle.

In the figure, the trigonometric functions have the following values:

$$\sin\theta = \frac{\text{opposite side}}{\text{hypotenuse}} = \frac{y}{r},$$

$$\cos\theta = \frac{\text{adjacent side}}{\text{hypotenuse}} = \frac{x}{r},$$

$$\tan\theta = \frac{\text{opposite side}}{\text{adjacent side}} = \frac{y}{x},$$

$$\cot\theta = \frac{1}{\tan\theta} = \frac{x}{y},$$

$$\sec\theta = \frac{1}{\cos\theta} = \frac{1}{\frac{x}{r}} = \frac{r}{x}, \quad \text{and}$$

$$\csc\theta = \frac{1}{\sin\theta} = \frac{1}{\frac{y}{r}} = \frac{r}{y}.$$

Also, from the figure, $r^2 = x^2 + y^2$ (from the Pythagorean Theorem), or $r = \sqrt{x^2 + y^2}$. Therefore, in this problem,

$$\tan\theta = 2 = \frac{-2}{-1} = \frac{y}{x}.$$

Hence, because of similar triangles, we can choose $y = -2$ and $x = -1$. Also, in this problem,

$$r^2 = x^2 + y^2 = (-1)^2 + (-2)^2 = 1 + 4$$

or $r^2 = 5$ or $r = \sqrt{5}$. Therefore,

$$\sin\theta = \frac{-2}{\sqrt{5}} = -\frac{2\sqrt{5}}{5}, \qquad \cos\theta = \frac{-1}{\sqrt{5}} = -\frac{\sqrt{5}}{5},$$

$$\tan\theta = 2, \qquad \cot\theta = \frac{-1}{-2} = \frac{1}{2},$$

$$\sec\theta = \frac{\sqrt{5}}{-1} = -\sqrt{5}, \quad \text{and} \quad \csc\theta = \frac{\sqrt{5}}{-2} = -\frac{\sqrt{5}}{2}.$$

 Find all the sides and angles of triangle ABC, given $a = 137$, $c = 78.0$, $\angle C = 23°0'$.

 Draw triangle ABC, filling in the given information. Thus

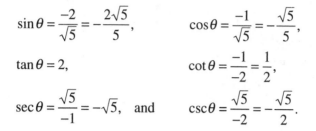

we divide the problem into 3 parts:

(1) find angle A.

(2) find angle B.

(3) find side b.

In order to find angle A we may use the law of sines, $\dfrac{\sin A}{\text{side } a} = \dfrac{\sin C}{\text{side } c}$, because we are given side $a = 137$, side $c = 78$, and $\sin C = \sin 23°$; thus, $\dfrac{\sin A}{137} = \dfrac{\sin 23°0'}{78}$.

Using our Trig. table we find $\sin 23° = 0.3907$. Thus

$$\frac{\sin A}{137} = \frac{0.3907}{78}$$

Multiplying both sides by 137 we obtain

$$\sin A = \frac{137(0.3907)}{78}$$

$$\sin A = 0.6862.$$

Using our Trig. table, we find that $\angle A = 43°20'$. Our Trig. table only gives values of sine between $0°$ and $90°$. Since we are dealing with an angle in a triangle, which can take on values greater than $90°$ (recall there are $180°$ in a triangle), we must examine what happens to the sine function in the second quadrant, that is between $90°$ and $180°$.

We use our trigonometric identity $\sin\theta = \sin(180 - \theta)$:

$$\sin 43°20' = \sin(180 - 43°20') = \sin 136°40'$$

Hence $\angle A = 43°20'$ *or* $136°40'$

Thus there are two solutions. We now proceed to the next part of our problem, finding angle B, and side b.

$$\angle A + \angle B + C = 180° \quad \text{(There are } 180° \text{ in a triangle)}$$

$$\angle A + \angle B + 23° = 180°$$

$$\angle A + \angle B = 157°$$

If $\angle A = 43°20'$ then

$$43°20' + \angle B = 157°$$

$$\angle B = 157° - 43°20' = 113°40'$$

Since we now know $\angle B$, we may apply the law of sines:

$$\frac{\sin B}{\text{side } b} = \frac{\sin C}{\text{side } c}$$

to find side b:

$$\frac{\sin 113°40'}{b} = \frac{\sin 23°}{78}$$

Cross multiplying we obtain

$$b \sin 23° = 78 \sin 113°40'$$

$$b = \frac{78 \sin 113°40'}{\sin 23°}$$

Substituting in the values sin 113°40' = 0.9159 and sin 23° = 0.3907 we obtain

$$b = 78 \left(\frac{0.9159}{0.3907} \right) = 183$$

Hence if we choose ∠A = 43°20' then

$$\angle B = 113°40' \text{ and}$$

$$\text{side } b = 183$$

If, however, we choose ∠A = 136°40', then since

$$\angle A + \angle B = 157°$$

$$136°40' + \angle B = 157°$$

and $\angle B = 157° - 136°40' = 20°20'$

Applying the law of sines to find side b gives us:

$$\frac{\sin B}{\text{side } b} = \frac{\sin C}{\text{side } c}$$

$$\frac{\sin 20°20'}{b} = \frac{\sin 23°}{78}$$

Cross multiplying gives us

$$b \sin 23° = 78 \sin 20°20'$$

$$b = \frac{78 \sin 20°20'}{\sin 23°}$$

Substituting in sin 20°20' = 0.3475 and sin 23° = 0.3907 we obtain

$$b = \frac{78(0.3475)}{0.3907} = 69$$

Hence if we choose $\angle A = 136°40'$, then $\angle B = 20°20'$, and side b = 69.

 Given $a = 8$, $c = 7$, $\beta = 135°$, find b.

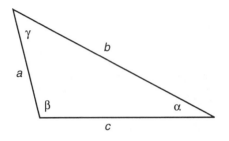

 Use the law of cosines to find one side given two sides and an included angle.

$b^2 = a^2 + c^2 - 2ac \cos$ (included angle) where the included angle is the angle between the two given sides.

$$b^2 = a^2 + c^2 - 2ac \cos \beta$$
$$b^2 = 64 + 49 - 2 \times 8 \times 7 \cos 135°$$
$$= 113 - 112 \times \left(\frac{-\sqrt{2}}{2} \right) = 113 + 79.195$$
$$= 192.195$$
$$b = 13.863$$

6.5 Solving Trigonometric Equations

The procedures for solving trigonometric equations are similar to the procedures for solving algebraic equations. However, it is often desirable to use substitutions, so that the equation being solved contains only one of the trigonometric functions. The example below illustrates this technique.

EXAMPLE

Solve:

$\sin x + 1 = 2 \cos^2 x$ for $0 \le x < 360°$.

$\sin x + 1 = 2(1 - \sin^2 x)$

$\sin x + 1 = 2 - 2 \sin^2 x$

$2 \sin^2 x + \sin x - 1 = 0$

$(2 \sin x - 1)(\sin x + 1) = 0$

$2 \sin x - 1 = 0$ or $\sin x + 1 = 0$

$\sin x = \frac{1}{2}$ or $\sin x = -1$

$x = 30°, 150°, 270°.$

Problem Solving Examples:

 Find all values of θ such that $\sec \theta = 2$.

 First find all values of θ, $0° \le \theta < 360°$, such that $\sec \theta = 2$. They are $\theta = 60°$ and $\theta = 300°$.

To obtain all values of θ for which $\sec \theta = 2$, add multiples of $360°$(or $2n\pi$) where $n = 0, \pm 1, \pm 2, \pm 3....$

Hence, all the values of θ for which $\sec \theta = 2$ are given by

$\theta = 60° + 360° \times n$ and $\theta = 300° + 360° \times n$

(or $\theta = \frac{\pi}{3} + 2n\pi$ and $\theta = \frac{5}{3}\pi + 2n\pi$),

where n is any integer.

Find the solution set on $[0, 2\pi]$ of the equation

$$\sqrt{1 + \sin^2 x} = \sqrt{2} \sin x.$$

Since the unknown quantity is involved in the radicand, squaring of both sides to eliminate the radical is suggested. Thus, we obtain

$1 + \sin^2 x = 2\sin^2 x$. Hence, $\sin^2 x - 1 = 0$,

$\sin^2 x = 1$

or $\sqrt{\sin^2 x} = \pm\sqrt{1}$

$\sin x = \pm 1$

When $\sin x = 1$ on $[0, 2\pi]$, $x = \dfrac{\pi}{2}$. When $\sin x = -1$ on $[0, 2\pi]$, $x = \dfrac{3\pi}{2}$.

The complete solution set seems to be $\{\dfrac{\pi}{2}, \dfrac{3\pi}{2}\}$. Since we squared both sides of the equation, we should try each element in the original equation. When $x = \dfrac{\pi}{2}$, we obtain $\sqrt{1+1} = \sqrt{2} \times 1$ When $x = \dfrac{3\pi}{2}$, we obtain $\sqrt{1+1} = \sqrt{2}(-1)$. The second element does not satisfy the original equation, hence does not belong to the solution set. An extraneous root was introduced by squaring the equation. Thus, the solution set is $\{\dfrac{\pi}{2}\}$.

Solve $2\sin^2\theta + 3\cos\theta - 3 = 0$ for θ if $0 \le \theta < 360°$.

The solution to the equation can be found by expressing the equation in terms of one trigonometric function. Here the convenient function is cos θ. Using the identity $\sin^2\theta + \cos^2\theta = 1$, we can eliminate $\sin^2\theta$ from the equation by substituting $1 - \cos^2\theta$:

$$2\left(1 - \cos^2\theta\right) + 3\cos\theta - 3 = 0$$

distributing: $2 - 2\cos^2 + 3\cos\theta - 3 = 0$

adding: $-2\cos^2\theta + 3\cos\theta - 1 = 0$

multiply by -1: $2\cos^2\theta - 3\cos\theta + 1 = 0$

factoring: $(2\cos\theta - 1)(\cos\theta - 1) = 0$

Hence, $\cos\theta = \dfrac{1}{2}$ or $\cos\theta = 1$

From the 30°–60° right triangle on page 175 we find

$$\cos 60° = \frac{\text{adjacent side}}{\text{hypotenuse}} = \frac{1}{2}.$$

The cosine is positive in quadrants I and IV; thus, θ can be 60° or 300°. Then,

$$\theta = 60°, 300° \quad \text{or} \quad \theta = 0°(\text{since } \cos 0° = 1)$$

Substitution verifies all three. The solution set is

$$\{0°, 60°, 300°\}$$

By removing the restriction that $0 \le \theta < 360°$, the solution set is $\{0° + 360°k, 60° + 360°k, 300° + 360°k\}$, k an integer.

 Find the solution set of $2\cos^2 x - 5\cos x + 2 = 0$.

Factoring, we obtain $(\cos x - 2)(2\cos x - 1) = 0$. Setting each factor equal to zero, we obtain $\cos x = 2$ and $\cos x = 1/2$. There is no value of x satisfying the first factor because the range of values for $\cos x$ is from -1 to $+1$; that is, $-1 \le \cos x \le 1$. Therefore, the solution set of the first factor is the empty set. For the second factor the solution set is

$$\left\{ x \middle| x = \frac{\pi}{3} + 2n\pi \ \text{or} \ \frac{5\pi}{3} + 2n\pi, \ n = 0, \pm 1, \pm 2, ... \right\}.$$

Since the first solution set is the empty set, the second set is the complete solution set.

6.6 The Trigonometric Form of a Complex Number

In an earlier section, it was determined that when a and b are real numbers, $a + bi$ is a complex number. Then, it is natural to associate $a + bi$ with the point (a, b). It is obvious that this procedure establishes a one-to-one correspondence between the set of all complex numbers and the set of all points in a plane.

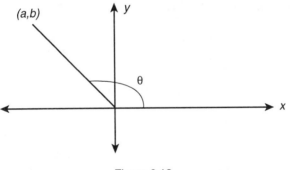

Figure 6.12

A segment is drawn from the origin to (a, b). Then, θ is the angle formed by this segment and the positive x axis, and r is the distance from the origin to (a, b). If $z = a + bi$, then

$Z = r(\cos \theta + i \sin \theta)$

and $r(\cos \theta + i \sin \theta)$ is called the trigonometric form of Z.

Products and quotients of two complex numbers can be determined by the formulas below.

If $Z_1 = r_1(\cos \theta_1 + i \sin \theta_1)$ and $Z_2 = r_2(\cos \theta_2 + i \sin \theta_2)$

then $Z_1 Z_2 = r_1 r_2 [\cos(\theta_1 + \theta_2) + i \sin (\theta_1 + \theta_2)]$

and $\dfrac{Z_1}{Z_2} = \dfrac{r_1}{r_2}\left[\cos(\theta_1 - \theta_2) + i\sin(\theta_1 - \theta_2)\right]$ $r_2 \neq 0$

The product rule extends to n factors, and when the n factors are identical, and if $Z = r(\cos \theta + i \sin \theta)$, then

$$Z^n = r^n (\cos n\theta + i \sin n\theta).$$

This is called De Moivre's Theorem.

Suppose $Z = r(\cos\theta + i\sin\theta)$. Then applying De Moivre's Theorem to

$$\left[r^{\frac{1}{n}}\left(\cos\frac{\theta}{n} + i\sin\frac{\theta}{n} \right) \right]^n,$$

$$\left[r^{\frac{1}{n}}\left(\cos\frac{\theta}{n} + i\sin\frac{\theta}{n} \right) \right]^n = \left(r^{\frac{1}{n}} \right)^n \left(\cos\frac{\theta}{n}n + i\sin\frac{\theta}{n}n \right)$$

$$= r(\cos\theta + i\sin\theta)$$

$$= Z$$

It follows that

$$r^{\frac{1}{n}}\left(\cos\frac{\theta}{n} + i\sin\frac{\theta}{n} \right)$$

is an nth root of Z. Also, since

$$\cos\theta = \cos(\theta + K \times 360°) \quad \text{and} \quad \sin\theta = \sin(\theta + K \times 360°),$$

it is possible to find the n, n^{th} roots of Z. This process is illustrated in the following example.

EXAMPLE

Find the four fourth roots of $2 + 2\sqrt{3}\, i$

$Z = 4(\cos 60° + i\sin 60°),$

$Z = 4(\cos 420° + i\sin 420°)$

$Z = 4(\cos 780° + i\sin 780°),$

$Z = 4(\cos 1140° + i\sin 1140°)$

$W_1 = 4^{\frac{1}{4}}\left(\cos\frac{60°}{4} + i\sin\frac{60°}{4}\right) = \sqrt{2}(\cos 15° + i\sin 15°)$

$W_2 = 4^{\frac{1}{4}}\left(\cos\frac{420°}{4} + i\sin\frac{420°}{4}\right) = \sqrt{2}(\cos 105° + i\sin 105°)$

$W_3 = 4^{\frac{1}{4}}\left(\cos\frac{780°}{4} + i\sin\frac{780°}{4}\right) = \sqrt{2}(\cos 195° + i\sin 195°)$

$W_4 = 4^{\frac{1}{4}}\left(\cos\frac{1140°}{4} + i\sin\frac{1140°}{4}\right) = \sqrt{2}(\cos 285° + i\sin 285°)$

Problem Solving Examples:

 Express each of the following in trigonometric form.

(a) $-\sqrt{2} + \sqrt{2}i$ (b) $3 - 4i$ (c) $2 + i$

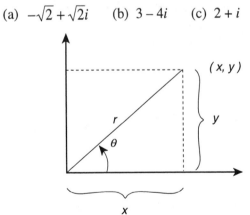

 In the plane, a complex number is represented as $x + iy$.

Therefore the angle θ can be defined as

$$\arctan \frac{y}{x} = \theta \quad \text{or} \quad \tan \theta = \frac{y}{x}.$$

In part (a) the x coordinate is negative and the y coordinate is positive, therefore θ must lie in the second quadrant.

In part (b) the x is positive and the y is negative so θ lies in the fourth quadrant.

Finally in part (c) both x and y are positive implying that θ lies in the first quadrant.

The modulus or radius can be computed from the Pythagorean theorem $r^2 = x^2 + y^2$.

(a) $\tan \theta = \dfrac{\sqrt{2}}{-\sqrt{2}} = -1$, and θ is in the second quadrant.

Since $\arctan 1 = 45°$, $\theta = 180° - 45° = 135°$.

$$r^2 = \left(-\sqrt{2}\right)^2 + \left(\sqrt{2}\right)^2 = 4 \quad \text{or} \quad r = \sqrt{4} = 2.$$

Therefore, $-\sqrt{2} + \sqrt{2}i = 2 \left(\cos 135° + i\sin 135°\right).$

(b) $\tan \theta = \dfrac{-4}{3}$, and θ is in the fourth quadrant.

Since $\arctan 1.333 = 53° \ 10'$ (to the nearest 10'),

$$\theta = 360° - 53°10' = 306°50'$$

$$r^2 = 3^2 + (-4)^2 = 25 \quad \text{or} \quad r = 5.$$

Therefore, $3 - 4i = 5 (\cos 306°50' + i \sin 306°50').$

(c) $\tan \theta = \dfrac{1}{2}$, and θ is in the first quadrant.

Since $\arctan \dfrac{1}{2} = 26°30'$ (to the nearest 10'),

$$\theta = 26°30'$$

$$r^2 = 2^2 + 1^2 = 5 \quad \text{or} \quad r = \sqrt{5}.$$

Therefore, $2 + i = \sqrt{5}(\cos 26°30' + i \sin 26°30')$.

Q Find the value of $(4 - 4i) \cdot (\sqrt{3} - i)$ in polar form.

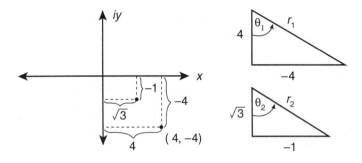

A First change each Cartesian representation to its polar representation. That is, we want to transform $x + iy$ to the form $r(\cos \theta + i \sin \theta)$. In the figure, notice that r can be determined using the Pythagorean theorem. And θ can be computed using the trigonometric functions.

$$r_1^2 = 4^2 + (-4)^2 = 32$$
$$r_1 = \sqrt{32} = 4\sqrt{2}$$
$$r_2^2 = (\sqrt{3})^2 + (-1)^2 = 3 + 1 = 4$$
$$r_2 = 2$$

Since all three sides of both triangles are known, any of the trigonometric functions can be used to determine θ. i.e. $\sin \theta = \dfrac{y}{r}$; $\tan \theta = \dfrac{y}{x}$.

Once θ_1 and θ_2 have been determined, the multiplication can be performed according to the formula

$$r_1(\cos\theta_1 + i\sin\theta_1) \times r_2(\cos\theta_2 + i\sin\theta_2)$$

$$= r_1 r_2(\cos[\theta_1 + \theta_2] + i\sin[\theta_1 + \theta_2]).$$

$$\tan\theta_1 = \frac{-4}{4} \to \theta_1 = \tan^{-1}(-1) = 315°, \quad \text{fourth quadrant.}$$

$$\tan\theta_2 = \frac{-1}{\sqrt{3}} \to \theta_2 = \tan^{-1}(-\sqrt{3}) = 330°, \quad \text{fourth quadrant.}$$

Changing $4 - 4i$ and $\sqrt{3} - i$ to polar form, we obtain

$$4 - 4i = 4\sqrt{2}(\cos 315° + i\sin 315°) \quad \text{and}$$

$$\sqrt{3} - i = 2(\cos 330° + i\sin 330°).$$

Thus,

$$(4 - 4i)(\sqrt{3} - i) = 4\sqrt{2}(\cos 315° + i\sin 315°) \times 2(\cos 330° + i\sin 330°)$$

$$= 8\sqrt{2}(\cos 645° + i\sin 645°) = 8\sqrt{2}(\cos 285° + i\sin 285°).$$

 Find $\left(8\left(\cos\dfrac{\pi}{2} + i\sin\dfrac{\pi}{2}\right)\right) \div \left(2\left(\cos\dfrac{\pi}{6} + i\sin\dfrac{\pi}{6}\right)\right).$

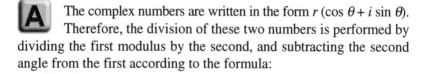

The complex numbers are written in the form $r(\cos\theta + i\sin\theta)$. Therefore, the division of these two numbers is performed by dividing the first modulus by the second, and subtracting the second angle from the first according to the formula:

$$\frac{r_1\left(\cos\theta_1 + i\sin\theta_1\right)}{r_2\left(\cos\theta_2 + i\sin\theta_2\right)}$$

$$= \frac{r_1}{r_2}\left[\cos\left(\theta_1 - \theta_2\right) + i\sin\left(\theta_1 - \theta_2\right)\right]$$

$$r_1 = 8$$

$$r_2 = 2$$

$$\frac{r_1}{r_2} = \frac{8}{2} = 4$$

$$\theta_1 = \pi/2$$

$$\theta_2 = \pi/6$$

$$\theta_1 - \theta_2 = \pi/2 - \pi/6 = \pi/3 = \text{argument}$$

$$8\left(\cos\frac{\pi}{2} + i\sin\frac{\pi}{2}\right) \div 2\left(\cos\frac{\pi}{6} + i\sin\frac{\pi}{6}\right)$$

$$= 4\left(\cos\frac{\pi}{3} + i\sin\frac{\pi}{3}\right)$$

Check

$$8\left(\cos\frac{\pi}{2} + i\sin\frac{\pi}{2}\right) = 8(0 + i) = 8i.$$

$$2\left(\cos\frac{\pi}{6} + i\sin\frac{\pi}{6}\right) = 2\left(\frac{\sqrt{3}}{2} + \frac{1}{2}i\right) = \sqrt{3} + i$$

$$\frac{8i}{\sqrt{3} + i} = \frac{8i}{\sqrt{3} + i} \cdot \frac{\sqrt{3} - i}{\sqrt{3} - i} = \frac{8 + 8\sqrt{3}i}{3 - i^2}$$

$$= \frac{8 + 8\sqrt{3}i}{4} = 2 + 2\sqrt{3}i$$

$$4\left(\cos\frac{\pi}{3}+i\sin\frac{\pi}{3}\right)=4\left(\frac{1}{2}+\frac{\sqrt{3}}{2}i\right)=2+2\sqrt{3}i$$

 Compute $\left(\cos\frac{3\pi}{2}+i\sin\frac{3\pi}{2}\right)^{6}$.

 To raise the trigonometric representation of a complex number to a power, apply the rule:

$$w = r(\cos\theta + i\sin\theta)$$
$$w^{n} = \left[r(\cos\theta + i\sin\theta)\right]^{n} = r^{n}(\cos n\theta + i\sin n\theta);$$

here $n = 6$, $r = 1$. Thus,

$$\left(\cos\frac{3\pi}{2}+i\sin\frac{3\pi}{2}\right)^{6} = \cos\frac{18\pi}{2}+i\sin\frac{18\pi}{2}$$
$$= \cos 9\pi + i\sin 9\pi.$$

Recall the formula for determining coterminal numbers, $u = u \pm 2n\pi$; $u = 9\pi$, $n = 4$. Then, $9\pi = 9\pi - 2(4)\pi = 9\pi - 8\pi = \pi$. This means that on the unit circle both π and 9π begin at $(1,0)$ and terminate at π. Thus, we have: $\cos\pi + i\sin\pi$, and since $\cos\pi = -1$ and $\sin\pi = 0$, $=$ $-1 + i(0) = -1$. Note that $\cos\dfrac{3\pi}{2} + i\sin\dfrac{3\pi}{2}$ is a number which can be raised to an even power to produce a negative product.

Find the 5th roots of $-1 + i$.

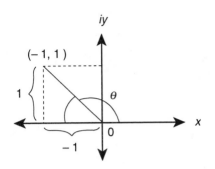

In the figure notice that we can determine r and θ so that we can transform the complex number from Cartesian coordinates to a polar representation. Use the Pythagorean theorem to determine r.

$$r^2 = x^2 + y^2 = (-1)^2 + (1)^2 = 2$$
$$r = \sqrt{2}$$

θ can be computed by using the fact that $\tan \theta = y/x$ or

$$\arctan \frac{y}{x} = \theta$$

$$\theta = \arctan\left(\frac{1}{-1}\right) = \arctan(-1)$$

The angle whose sin is $\dfrac{\sqrt{2}}{2}$ and whose cos is $\dfrac{-\sqrt{2}}{2}$ is $135°$

measured counterclockwise from $0°$.

The 5th roots are calculated according to the rule

$$w^{1/n} = r^{1/n}\left(\cos\frac{\theta + 2k\pi}{n} + i\sin\frac{\theta + 2k\pi}{n}\right)$$

where $0 \le k \le n-1$ inclusive

$$-1+i = \sqrt{2}(\cos 135° + i \sin 135°)$$

$$\sqrt[5]{-1+i} = \left(2^{\frac{1}{2}}\right)^{\frac{1}{5}}\left(\cos\frac{135° + 2k\pi}{5} + i \sin\frac{135° + 2k\pi}{5}\right)$$

$$k = 0, 1, \ldots, 4$$

Thus,

$$w_0 = 2^{\frac{1}{10}}(\cos 27° + i \sin 27°)$$
$$w_1 = 2^{\frac{1}{10}}(\cos 99° + i \sin 99°)$$
$$w_2 = 2^{\frac{1}{10}}(\cos 171° + i \sin 171°)$$
$$w_3 = 2^{\frac{1}{10}}(\cos 243° + i \sin 243°)$$
$$w_4 = 2^{\frac{1}{10}}(\cos 315° + i \sin 315°)$$

Quiz: Trigonometry

1. If it is known that $\sin\alpha = \dfrac{3}{4}$, what is $\cos\alpha$?

(A) $\dfrac{4}{\sqrt{7}}$.

(D) $\dfrac{4}{\sqrt{5}}$.

(B) $\dfrac{\sqrt{5}}{4}$.

(E) $\dfrac{\sqrt{7}}{4}$.

(C) $\dfrac{4}{3}$.

2. Evaluate $\tan^{-1}(-1)$

(A) $\dfrac{3\pi}{4}$.

(D) $\dfrac{-\pi}{4}$.

(B) $\dfrac{7\pi}{4}$.

(E) $\dfrac{-7\pi}{4}$.

(C) $\dfrac{-3\pi}{4}$.

3. What is the exact value of $\sin\left(\dfrac{7\pi}{12}\right)$?

(A) $\dfrac{2\sqrt{2}+\sqrt{6}}{4}$.

(D) $\dfrac{\sqrt{6}+\sqrt{2}}{4}$.

(B) $\dfrac{\sqrt{6}+\sqrt{3}}{2}$.

(E) $\dfrac{2+2\sqrt{3}}{3}$.

(C) $\dfrac{2+\sqrt{3}}{\sqrt{2}}$.

4. What is the exact value of $\sin\left(\arccos\dfrac{\sqrt{3}}{4}\right)$?

(A) $\dfrac{\sqrt{19}}{4}$.

(D) $\dfrac{\sqrt{3}}{4}$.

(B) $\dfrac{\sqrt{13}}{4}$.

(E) $\dfrac{13}{4}$.

(C) $\dfrac{2}{3}$.

5.   It is known that triangle ABC has $\angle A = 43°$, $\angle B = 37°$ and $a = 24$. What is the length of side c?

  (A) 23.64.   (D) 21.18.

  (B) 20.00.   (E) 35.67.

  (C) 34.66.

6.   Find all values of θ on $[0,2\pi]$ that satisfy the equation $\sin 2\theta = \cos \theta$.

  (A) $\dfrac{\pi}{4}$.   (D) $\dfrac{\pi}{6}, \dfrac{\pi}{2}$.

  (B) $\dfrac{\pi}{6}, \dfrac{\pi}{2}, \dfrac{5\pi}{6}, \dfrac{3\pi}{2}$.   (E) No solution.

  (C) $\dfrac{\pi}{7}, \dfrac{8\pi}{7}$.

7.   Let $z_1 = 1 + i$ and $z_2 = \sqrt{3} + i$. What is $\dfrac{z_1}{z_2}$ in trigonometric form?

  (A) $2\sqrt{2}(\cos 75° + i \sin 75°)$

  (B) $\sqrt{2}(\cos 345° + i \sin 345°)$

  (C) $\dfrac{1}{\sqrt{3}}(\cos 75° + i \sin 75°)$

  (D) $\dfrac{\sqrt{2}}{2}(\cos 15° + i \sin 15°)$

(E) $\dfrac{\sqrt{3}}{2}(\cos 15° + i\sin 15°)$

8. Which of the following is one of the 5th roots of $\sqrt{5} + \sqrt{5}i$?

(A) $5^{\frac{1}{10}}(\cos 9° + i\sin 9°)$

(B) $10^{\frac{1}{10}}(\cos 299° + i\sin 299°)$

(C) $10^{\frac{1}{5}}(\cos 81° + i\sin 81°)$

(D) $5^{\frac{1}{5}}(\cos 225° + i\sin 225°)$

(E) $10^{\frac{1}{10}}(\cos 153° + i\sin 153°)$

9. If α and β are acute angles such that $\sec\alpha = \dfrac{17}{15}$ and $\cos\beta = \dfrac{6}{10}$, find $\sin(\alpha+\beta)$.

(A) $\dfrac{84}{85}$ (D) $\dfrac{37}{38}$

(B) $\dfrac{56}{57}$ (E) $\dfrac{99}{100}$

(C) $\dfrac{7}{8}$

10. Determine which graph corresponds to $y = 3 + \sin 2x$.

(A)

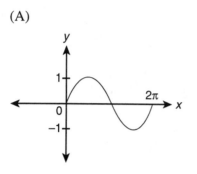

(B)

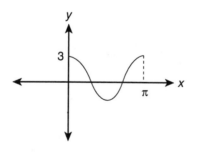

(C)

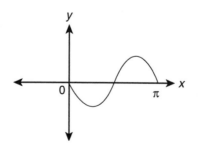

(D)

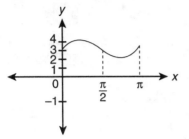

(E)

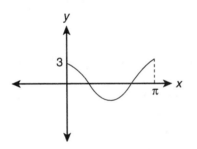

ANSWER KEY

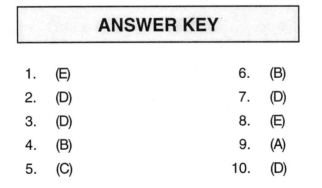

1.	(E)	6.	(B)
2.	(D)	7.	(D)
3.	(D)	8.	(E)
4.	(B)	9.	(A)
5.	(C)	10.	(D)

CHAPTER 7

Exponents and Logarithms

7.1 Integer Exponents

This section contains a summary of important interpretations of exponents. Here is a definition which describes the meaning of integer exponents.

If n is a positive integer, and if x is a real number, then:

$$x^n = \underbrace{x \times x \times x \times x \times x \times x \dots x}_{n \text{ factors}}$$

$$x^0 = 1 \quad \text{and}$$

$$x^{-n} = \frac{1}{x^n} \quad x \neq 0.$$

From these definitions, the following exponent generalizations can be derived.

If a and b are real numbers, and if m and n are integers, then:

$$a^m \times a^n = a^{m+n}$$

$$\frac{a^m}{a^n} = a^{m-n} \qquad a \neq 0$$

$$\left(a^{m}\right)^{n} = a^{mn}$$

$$(ab)^{n} = a^{n}b^{n}$$

$$\left(\frac{a}{b}\right)^{n} = \frac{a^{n}}{b^{n}} \qquad b \neq 0$$

Problem Solving Examples:

 Express $\left(\dfrac{a^{-2}}{b^{-3}}\right)^{-2}$ using only positive exponents.

 Solution A

By the law of exponents which states that $(x)^{-n} = \dfrac{1}{x^{n}}$ where n is a positive integer,

$$\left(\frac{a^{-2}}{b^{-3}}\right)^{-2} = \frac{1}{\left(\dfrac{a^{-2}}{b^{-3}}\right)^{2}}.$$

Since $\left(\dfrac{x}{y}\right)^{n} = \dfrac{x^{n}}{y^{n}}, \left(\dfrac{a^{-2}}{b^{-3}}\right)^{2} = \dfrac{\left(a^{-2}\right)^{2}}{\left(b^{-3}\right)^{2}}$. Also, since $\left(x^{m}\right)^{n} = x^{mn}$,

$$\left(a^{-2}\right)^{2} = a^{(-2)(2)} = a^{-4}, \left(b^{-3}\right)^{2} = b^{(-3)(2)} = b^{-6}. \text{ Hence,}$$

$$\left(\frac{a^{-2}}{b^{-3}}\right)^{-2} = \frac{1}{\left(\dfrac{a^{-2}}{b^{-3}}\right)^{2}}$$

$$= \frac{1}{\dfrac{a^{-4}}{b^{-6}}}$$

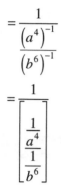

$$= \frac{1}{\dfrac{\left(a^4\right)^{-1}}{\left(b^6\right)^{-1}}}$$

$$= \frac{1}{\left[\dfrac{\dfrac{1}{a^4}}{\dfrac{1}{b^6}}\right]}$$

Note that division is the same as multiplying the numerator by the reciprocal of the denominator. This principle is applied to the term in brackets.

$$\left(\frac{a^{-2}}{b^{-3}}\right)^{-2} = \frac{1}{(1)\left(\dfrac{b^6}{a^4}\right)} = \frac{1}{\left(\dfrac{b^6}{a^4}\right)}.$$

Applying the same principle to the term in parenthesis on the right side of the equation:

$$\left(\frac{a^{-2}}{b^{-3}}\right)^{-2} = \left(\frac{\dfrac{1}{b^6}}{a^4}\right) = (1)\left(\frac{a^4}{b^6}\right) = \frac{a^4}{b^6}.$$

Solution B   Since $\left(\dfrac{x}{y}\right)^n = \dfrac{x^n}{y^n}, \left(\dfrac{a^{-2}}{b^{-3}}\right)^{-2} = \dfrac{\left(a^{-2}\right)^{-2}}{\left(b^{-3}\right)^{-2}}.$

Also, since $\left(x^m\right)^n = x^{mn}, \left(a^{-2}\right)^{-2} = a^{(-2)(-2)} = a^4,$

and $\left(b^{-3}\right)^{-2} = b^{(-3)(-2)} = b^6.$

Hence,

$$\left(\frac{a^{-2}}{b^{-3}}\right)^{-2} = \frac{a^4}{b^6}.$$

 Evaluate the following expression: $\dfrac{12x^7y}{3x^2y^3}$

 Noting (1) $\dfrac{abc}{def} = \dfrac{a}{d} \times \dfrac{b}{e} \times \dfrac{c}{f}$, (2) $a^{-b} = \dfrac{1}{a^b}$ and (3) $\dfrac{a^b}{a^c} = a^{b-c}$

for all non-zero values of a,d,e,f, we proceed to evaluate the expression:

$$\frac{12x^7y}{3x^2y^3} = \frac{12}{3} \times \frac{x^7}{x^2} \times \frac{y}{y^3} = 4 \times x^{7-2} \times y^{1-3} = 4x^5y^{-2} = \frac{4x^5}{y^2}.$$

Perform the indicated operations and simplify:

$$\left(\frac{-5b^y}{3^2x^5}\right)^3 \left(\frac{3x^7}{5b^y}\right)^2$$

A

$$\left(\frac{-5b^y}{3^2x^5}\right)^3 \left(\frac{3x^7}{5b^y}\right)^2 = \frac{\left(-5b^y\right)^3}{\left(3^2x^5\right)^3} \times \frac{\left(3x^7\right)^2}{\left(5b^y\right)^2} \quad \text{since} \quad \left(\frac{a}{b}\right)^x = \frac{a^x}{b^x}$$

$$= \frac{(-5)^3\left(b^y\right)^3}{\left(3^2\right)^3\left(x^5\right)^3} \times \frac{3^2\left(x^7\right)^2}{5^2\left(b^y\right)^2} \quad \text{since} \quad (ab)^x = a^xb^x$$

$$= \frac{-5^3 b^{3y}}{3^6 x^{15}} \times \frac{3^2 x^{14}}{5^2 b^{2y}} \quad \text{since} \quad \left(a^x\right)^y = a^{x \cdot y}$$

$$= \frac{\left(-5^3 b^{3y}\right)\left(3^2 x^{14}\right)}{\left(3^6 x^{15}\right)\left(5^2 b^{2y}\right)}$$

$$= \frac{\left(3^2 x^{14}\right)\left(-5^3 b^{3y}\right)}{\left(3^6 x^{15}\right)\left(5^2 b^{2y}\right)} \quad \text{using the commutative law of multiplication}$$

$$= \left(3^{2-6}\right)\left(x^{14-15}\right)\left[-\left(5^{3-2}\right)\left(b^{3y-2y}\right)\right] \quad \text{because}$$

$$\frac{x^a}{x^b} = x^{a-b},$$

$$= \left(3^{-4}\right)\left(x^{-1}\right)\left(-5^1\right)\left(b^y\right)$$

$$= \frac{-5b^y}{3^4 x} \quad \text{because} \quad x^{-a} = \frac{1}{x^a}$$

$$= \frac{-5b^y}{3 \times 3 \times 3 \times 3 x}$$

$$= \frac{-5b^y}{81x}$$

 Use the properties of exponents, to perform the indicated operations in

$$\left(2^3 x^4 5^2 y^7\right)^5.$$

 Since the product of several numbers raised to the same exponent equals the product of each number raised to that exponent (i.e., $(abcd)^x = a^x b^x c^x d^x$) we obtain,

$$\left(2^3 x^4 5^2 y^7\right)^5 = \left(2^3\right)^5 \left(x^4\right)^5 \left(5^2\right)^5 \left(y^7\right)^5.$$

Recall that $\left(x^a\right)^b = x^{a \times b}$; thus

$$\left(2^3 x^4 5^2 y^7\right)^5 = \left(2^{3\times5}\right)\left(x^{4\times5}\right)\left(5^{2\times5}\right)\left(y^{7\times5}\right)$$
$$= 2^{15} x^{20} 5^{10} y^{35}.$$

7.2 Rational Number Exponents

Radicals were discussed in Section 3.4, and since radicals involve rational number exponents, a short review may be desirable. Here is a useful definition.

Suppose that m and n are positive integers with $n > 1$, and suppose that a is a real number such that $\sqrt[n]{a}$ is a real number. Then

$$a^{\frac{m}{n}} = \left(\sqrt[n]{a}\right)^m$$

Of course $\sqrt[n]{a}$ is not a real number when n is even and a is negative.

With this definition, it is easy to see that the exponent laws listed in Section 7.1 can be extended from integers to positive rational numbers. Then it is quite easy to see that $a^{\frac{m}{n}} = \sqrt[n]{a^m}$. Also,

$$a^{-\frac{m}{n}} = \frac{1}{a^{\frac{m}{n}}}$$
$$= \frac{1}{\left(\sqrt[n]{a}\right)^m}$$

and this describes what happens when the exponent is a negative rational number. Of course, the exponent laws described in Section 7.1 carry over to rational number exponents.

Problem Solving Examples:

Determine whether each of the following expressions is true or false. If false, explain why.

(1) $x^4 \times x^6 = x^{24}$

(6) $\sqrt{25} = \pm 5$

(2) $\dfrac{a^6}{a^2} = a^3$

(7) $\sqrt{a+b} = \sqrt{a} + \sqrt{b}$

(3) $\left(y^4\right)^2 = y^6$

(8) $x^{\frac{2}{3}} = \left(\sqrt[2]{x}\right)^5$

(4) $\dfrac{a^4}{a^{-4}} = a^{4-4} = a^0 = 1$

(9) $\dfrac{1}{a^{-1} + b^{-1}} = a + b$

(5) $a^4 + a^6 = a^{10}$

(10) $(a+b)^{-1} = a^{-1} + b^{-1}$

 (1) False. The rule for multiplying exponential values is $a^m \times a^n = a^{m+n}$. Therefore, $x^4 \times x^6 = x^{10} \neq x^{24}$.

(2) False. The rule for dividing exponential values is

$$\dfrac{a^m}{a^n} = a^{m-n}. \text{ Therefore, } \dfrac{a^6}{a^2} = a^{6-2}$$
$$= a^4 \neq a^3.$$

(3) False. $\left(y^4\right)^2 = y^{4 \times 2} = y^8 \neq y^6$

(4) False. $\dfrac{a^4}{a^{-4}} = a^{4-(-4)} = a^8 \neq 1$

(5) False. In order to add exponential values, the two values must be raised to the same exponent. Therefore,

$$a^4 + a^6 \neq a^{10}$$

(6) True.

(7) False. Squaring both sides yields $a + b = a + 2\sqrt{a}\sqrt{b} + b$ or $0 = 2\sqrt{a}\sqrt{b}$ which is not generally true. So, $\sqrt{a+b} \neq \sqrt{a} + \sqrt{b}$

(8) False. $x^{\frac{2}{3}} = \sqrt[5]{x^2}$

(9) False. $\dfrac{1}{a^{-1}+b^{-1}} = \dfrac{1}{\dfrac{1}{a}+\dfrac{1}{b}} = \dfrac{1}{\dfrac{b+a}{ab}}$

$$= \dfrac{ab}{b+a} \neq a+b$$

(10) False $(a+b)^1 = \dfrac{1}{a+b} \neq a^{-1}+b^{-1}$

 Find the value of $\sqrt[4]{-64a^4}$.

 We can rewrite $\sqrt[4]{-64a^4}$

$$\sqrt[4]{-64a^4} = \sqrt[4]{(-1)2^6 a^4}$$

$$= \left[(-1)2^6 a^4\right]^{\frac{1}{4}}$$

$$= (-1)^{\frac{1}{4}}\left(2^6\right)^{\frac{1}{4}}\left(a^4\right)^{\frac{1}{4}}$$

$$= \pm i\, 2^{\frac{3}{2}} a$$

$$= \pm i\, 2\sqrt{2}\, a$$

$$= \pm 2a\sqrt{2}\, i$$

 Find the numerical value of each of the following.

(a) $8^{\frac{2}{3}}$ (b) $25^{\frac{3}{2}}$

(a) Since $x^{\frac{a}{b}} = \left(x^{\frac{1}{b}}\right)^{a}$, $8^{\frac{2}{3}} = \left(8^{\frac{1}{3}}\right)^{2} = \left(\sqrt[3]{8}\right)^{2} = (2)^{2} = 4$

(b) Similarly, $25^{\frac{3}{2}} = \left(25^{\frac{1}{2}}\right)^{3} = 5^{3} = 125$.

Find the product $\sqrt[4]{x^3 y} \times \sqrt[4]{xy^2}$ and simplify.

Note that $\sqrt[x]{a} \times \sqrt[x]{b} = \sqrt[x]{ab}$; thus,

$$\sqrt[4]{x^3 y} \times \sqrt[4]{xy^2} = \sqrt[4]{\left(x^3 y\right)\left(xy^2\right)}.$$

Recall that when multiplying, we add exponents; hence

$$\left(x^3 y^1\right)\left(x^1 y^2\right) = \left(x^{3+1} y^{1+2}\right), \text{ and}$$

we obtain,

$$= \sqrt[4]{x^4 y^3}$$

$$= \sqrt[4]{x^4}\left(\sqrt[4]{y^3}\right)$$

Now, since

$$\sqrt[4]{x^4} = \left(x^{\frac{1}{4}}\right)^4 = x^1 = x, \quad \sqrt[4]{x^3 y} \times \sqrt[4]{xy^2} = x\sqrt[4]{y^3}.$$

7.3 Exponential Functions

The previous section includes an interpretation of rational number exponents. Now, a pertinent question to ask is, "What happens when the exponent is an irrational number?" Since $\sqrt{2} \approx 1.4142$

$$2^{1.4} < 2^{\sqrt{2}} < 2^{1.5}$$

$$2^{1.41} < 2^{\sqrt{2}} < 2^{1.42}$$

$$2^{1.414} < 2^{\sqrt{2}} < 2^{1.415}$$

Using this process, it can be shown that $2^{\sqrt{2}}$ is in fact a single real number. This idea can be generalized, and if b is a real number, $b > 0$, b^x is a real number for every real number x. Then

$$f(x) = b^x$$

is called an exponential function. Here is the graph of $f(x) = 2^x$.

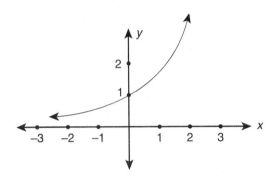

The irrational number e is defined in the following way.

$$\lim_{n \to \infty}\left(1 + \frac{1}{n}\right)^n = e$$

It can be shown that $e \approx 2.71828$.

Problem Solving Examples:

 Construct the graph of $y = 3^x$.

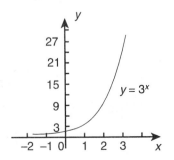

Assume values of x and compute the corresponding values of y by substituting into $y = 3^x$, obtaining the following table of values:

x:	-3	-2	-1	0	1	2	3
y:	$\dfrac{1}{27}$	$\dfrac{1}{9}$	$\dfrac{1}{3}$	1	3	9	27

The points corresponding to these pairs of values are plotted on the coordinate system of the figure and these points are joined by a smooth curve, which is the desired graph of the function. Note that the values of y are all positive. Furthermore, if $x < 0$, then y increases to a small extent as x does; if $x > 0$, y increases at a more rapid rate.

Q Graph the following functions:

(A) $y = 2^x$,

(B) $y = 4^x$,

(C) $y = 4^{-x}$,

(D) $y = 3 \times 2^x$.

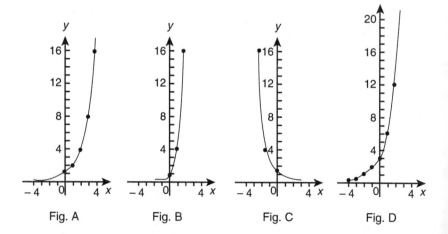

Fig. A   Fig. B   Fig. C   Fig. D

A When graphing a function $y = f(x)$, set up a table consisting of two columns: one for x and one for y. Choose values for x and find the corresponding y.

(A) $x = -4$, then $y = 2^x = 2^{-4} = \dfrac{1}{2^4} = \dfrac{1}{16}$.

Similarly, find other y values for different values of x. It is best to choose negative and positive values of x centering around and including zero to determine the nature of the graph.

x	-4	-3	-2	-1	0	1	2	3	4
y	$\dfrac{1}{16}$	$\dfrac{1}{8}$	$\dfrac{1}{4}$	$\dfrac{1}{2}$	1	2	4	8	16

Plot these points and draw a smooth curve through them. This is the graph of the exponential function $y = 2^x$.

Note that from the table and graph constructed, as you increase x by 1 each time moving from $x = -4$ to $x = 0$, the y values increase slightly. However, when you move through the positive values of x, the change in y is much greater for each unit change in x.

See Figure A.

(B) Construct a table in the same manner as problem A. The table of values for the integers -3 to 3 can be determined to be:

x	-3	-2	-1	0	1	2	3
y	$\dfrac{1}{64}$	$\dfrac{1}{16}$	$\dfrac{1}{4}$	1	4	16	64

Then plot these points and draw a smooth curve.

Figure B is the graph of $y = 4^x$, although it is not practical to plot the points corresponding to $x = -3$ or $x = 3$ on this coordinate system.

When this curve is compared to the graph of $y = 2^x$ (Figure A), we see that the general shape is the same. Both curves pass through the point $(0,1)$, that is, both have a y-intercept of 1. If we consider the curves to the left of the y-axes, we see that the curve of Figure B approaches the x-axis faster than the curve of Figure A. If we consider the same negative value on both curves, the y-value in Figure B is smaller than in Figure A. If

$$x = -3, \quad \text{then} \quad y = 2^{-3} = \frac{1}{8} \text{ for Figure A and}$$

$$y = 4^{-3} = \frac{1}{64} \text{ for Figure B. Thus the point } \left(-3, \frac{1}{64}\right)$$

is closer to the x-axis than $\left(-3, \frac{1}{8}\right)$.

(C) Obtain a table of ordered pairs as in Examples A and B. Plot the points and draw the smooth curve by connecting them.

$$f(-3) = 4^{-(-3)} = 4^3 = 64$$

$$f(-2) = 4^{-(-2)} = 4^2 = 16$$

$$f(-1) = 4^{-(-1)} = 4^1 = 4$$

$$f(0) = 4^0 = 1$$

$$f(1) = 4^{-1} = \frac{1}{4}$$

$$f(2) = 4^{-2} = \frac{1}{4^2} = \frac{1}{16}$$

$$f(3) = 4^{-3} = \frac{1}{4^3} = \frac{1}{64}$$

x	-3	-2	-1	0	1	2	3
y	64	16	4	1	$\frac{1}{4}$	$\frac{1}{16}$	$\frac{1}{64}$

Figure C is the graph of $y = 4^{-x}$.

The graph of $y = 4^x$ in Figure B and the graph of $y = 4^{-x}$ in Figure C are mirror images of each other.

(D) In Example A we determined the values of 2^x for x an integer and $-4 \leq x \leq 4$. The values of y for this function then must be three times the corresponding values of y of Example A.

x	-4	-3	-2	-1	0	1	2	3	4
y	$\dfrac{3}{16}$	$\dfrac{3}{8}$	$\dfrac{3}{4}$	$\dfrac{3}{2}$	3	6	12	24	48

The graph of this function is shown in Figure D.

From these four examples we can see some of the features of the graph of $y = ab^x$, $a > 0$, and $b > 0$. The y-intercept of the function is a: If $b > 1$, the curve will approach the x-axis to the left of the y-axis and the y value increases as the x value increases. The graph will be in quadrants I and II.

The y-intercept of Examples A, B, and C is 1 since $a = 1$. It is true for Examples A and B that the curve approaches the x-axis as x becomes more negative and the y-value increases as x increases. However, when the function contains a negative exponential expression in Example C, the reverse occurs. As x decreases, y increases and the curve approaches the x-axis as x becomes more positive.

7.4   Logarithms

A logarithm is an exponent. This is illustrated in the following definition.

For $b > 0$, $b \neq 1$ and for $x > 0$,

$y = \log_b x$ if and only if $b^y = x$.

Here are some examples which illustrate the use of this definition.

EXAMPLE

Find $\log_2 8$.

$$2^3 = 8$$

Thus, $\log_2 8 = 3$.

Find $\log_2 \dfrac{1}{2}$.

$$\frac{1}{2} = 2^{-1}$$

Thus, $\log_2 \dfrac{1}{2} = -1$.

Find $\log_2 \sqrt{2}$.

$$2^{\frac{1}{2}} = \sqrt{2}$$

Thus, $\log_2 \sqrt{2} = \dfrac{1}{2}$.

Find $\log_2(-2)$.

Since there does not exist a real number x such that

$$2^x = -2, \log_2(-2) \text{ is undefined.}$$

Since a logarithm is an exponent, it is easy to use exponent laws to establish the following generalizations.

If x_1, x_2, and b are positive real numbers, with $b \neq 1$, and m represents any real number,

$$\log_b(x_1 x_2) = \log_b x_1 + \log_b x_2$$

$$\log_b\left(\frac{x_1}{x_2}\right) = \log_b x_1 - \log_b x_2$$

$$\log_b x^m = m\log_b x \quad \text{and}$$

$$\log_b b^m = m.$$

The irrational number e was defined in Section 7.3. This number is often used as a base in the investigation of logarithms. A special symbology "ln" has been developed for this situation. Specifically,

$$\log_e a = \ln a.$$

The properties of exponents can be used in equation solving situations. Here are some examples.

EXAMPLE

Solve:

$$5 = 9e^{1-2x}$$
$$\ln 5 = \ln 9e^{1-2x}$$
$$\ln 5 = \ln 9 + \ln e^{1-2x}$$
$$\ln 5 = \ln 9 + 1 - 2x$$
$$2x = 1 + \ln 9 - \ln 5$$
$$x = \frac{1 + \ln 9 - \ln 5}{2}$$

Solve:

$$\log_3 x + \log_3(x+2) = 1$$
$$\log_3 x(x+2) = 1$$
$$x(x+2) = 3^1$$
$$x^2 + 2x - 3 = 0$$
$$(x+3)(x-1) = 0$$
$$x = -3 \quad \text{or} \quad x = 1$$

However, x cannot be a negative number. Thus, the only solution is $x = 1$.

Solve:

$$2^x = 5$$

$$\log_2 2^x = \log_2 5$$

$$x = \log_2 5$$

It is sometimes desirable to change from one logarithmic base to another. The following formula illustrates this process.

$$\log_a x = \frac{\log_b x}{\log_b a}$$

Sometimes if the base b is not important, or is understood, we write simply log x.

Problem Solving Examples:

Express the logarithm of $\dfrac{\sqrt{a^3}}{c^5 b^2}$ in terms of log a, log b and log c.

We apply the following properties of logarithms:

$$\log_b(P \times Q) = \log_b P + \log_b Q$$

$$\log_b(P/Q) = \log_b P - \log_b Q$$

$$\log_b\left(P^n\right) = n\log_b P$$

$$\log_b\left(\sqrt[n]{P}\right) = \frac{1}{n}\log_b P$$

Therefore,

$$\log \frac{\sqrt{a^3}}{c^5 b^2} = \log \frac{a^{\frac{3}{2}}}{c^5 b^2}$$

$$= \log a^{\frac{3}{2}} - \log\left(c^5 b^2\right)$$

$$= \frac{3}{2}\log a - \left(\log c^5 + \log b^2\right)$$

$$= \frac{3}{2}\log a - \log c^5 - \log b^2$$

$$= \frac{3}{2}\log a - 5\log c - 2\log b.$$

 Write the following equations in logarithmic form.

(a) $3^4 = 81$

(b) $10^0 = 1$

(c) $M^k = 5$

(d) $5^k = M$

 The expression $b^y = x$ is equivalent to the logarithmic expression $\log_b x = y$. Hence,

(a) $3^4 = 81$ is equivalent to the logarithmic expression $\log_3 81 = 4$.

(b) $10^0 = 1$ is equivalent to the logarithmic expression $\log_{10} 1 = 0$.

(c) $M^k = 5$ is equivalent to the logarithmic expression $\log_M 5 = k$.

(d) $5^k = M$ is equivalent to the logarithmic expression $\log_5 M = k$.

Solve $\log_2(x-1)+\log_2(x+1)=3$

 Applying a property of logarithms, $\log_b x + \log_b y = \log_b xy$, to

$\log_2(x-1) + \log_2(x+1) = 3$

we get $\log_2[(x-1)(x+1)]=3$. $\log_b x = y$ is equivalent to $b^y = x$ by definition, thus $\log_2[(x-1)(x+1)]=3$ is equivalent to

$$(x-1)(x+1) = 2^3 = 8$$
$$x^2 - 1 = 8$$
$$x^2 - 9 = 0$$
$$(x+3)(x-3) = 0.$$

$(x+3)(x-3)=0$ means either

$x+3=0$ or $x-3=0$

and

$x=-3$ or $x=3$.

Therefore, $\{3,-3\}$ is the possible solution set, but we must check each in the given equation. This is necessary because we have not defined the logarithm of a negative number and, consequently, must rule out any value of x which would require the use of the logarithm of a negative number.

Check: Replacing x with 3 in our original equation:

$\log_2(x-1)+\log_2(x+1)=3$
$\log_2(3-1)+\log_2(3+1)=3$
$\log_2 2 + \log_2 4 = 3$
$1+2=3$ since $2^1=2$ and $2^2=4$
$3=3$.

Replacing x by (-3) in our original equation

$$\log_2(x-1)+\log_2(x+1)=3$$
$$\log_2(-3-1)+\log_2(-3+1)=3$$
$$\log_2(-4)+\log_2(-2)=3.$$

$x = -3$ cannot be accepted as a root because we have not defined the logarithm of a negative number. Thus our solution set is {3}.

 Find $\log_{10}100$.

 The following solution presents two methods for solving the given problem.

Method I.

The statement $\log_{10}100 = x$ is equivalent to $10^x = 100$.

Since $10^2 = 100$, $\log_{10}100 = 2$.

Method II.

Note that $100 = 10 \times 10$; thus $\log_{10}100 = \log_{10}(10 \times 10)$. Recall: $\log^x(a + b) = \log^x a + \log^x b$, therefore

$$\log_{10}(10 \times 10) = \log_{10}10 + \log_{10}10$$
$$= 1 + 1$$
$$= 2.$$

 If $\log_3 N = 2$, find N.

 Since $\log_3 N = 2$ is equivalent to the equation $3^2 = N$, we obtain $N = 9$.

7.5 Logarithmic Functions

A function of the form

$$f(x) = \log_b x$$

is called a logarithmic function. In section 7.4, it was pointed out that $\log_b x$ is meaningful only when $b > 0$ and $b \neq 1$. Here are some properties of $f(x) = \log_b x$.

(1) The domain of f is the set of all positive real numbers.

(2) The range of f is the set of all real numbers.

(3) $f(1) = 0$

(4) f is an increasing function when $b > 1$ and decreasing if $0 < b < 1$.

From the definition of a logarithm, it is obvious that

$$f(x) = \log_b x \quad \text{and} \quad f(x) = b^x$$

are inverse functions. The graph of

$$f(x) = 2^x \quad \text{and} \quad f(x) = \log_2 x$$

is pictured below.

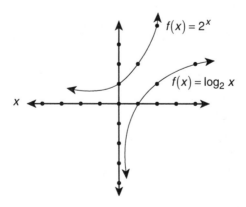

Construct the graph of $y = \log_2 x$.

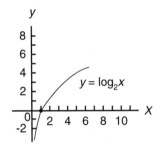

The equations $u = \log_b v$ and $v = b^u$ are equivalent.

Therefore, the relation $y = \log_2 x$ is equivalent to $x = 2^y$. Hence we assume values of y and compute the corresponding values of x, getting the table:

$x:$	$\frac{1}{8}$	$\frac{1}{4}$	$\frac{1}{2}$	1	2	4	8
$y:$	-3	-2	-1	0	1	2	3

For example, if $y = -3$, then $x = 2^y = 2^{-3} = \dfrac{1}{2^3} = \dfrac{1}{8}$.

The points corresponding to these values are plotted on the coordinate system in the figure. The smooth curve joining these points is the desired graph of $y = \log_2 x$. It should be noted that the graph lies entirely to the right of the y-axis. The graph of $y = \log_b x$ for any $b > 1$ will be similar to that in the figure. Some of the properties of this function which can be noted from the graph are:

I. $\log_b x$ is not defined for negative values of x or zero.

II. $\log_b 1 = 0$.

III. If $x > 1$, then $\log_b x > 0$.

IV. If $0 < x < 1$, then $\log_b x < 0$.

Q (a) Graph the functions

(1) $y = \ln x$

(2) $y = e^x$

(b) State the domain and range for (1) and (2).

A

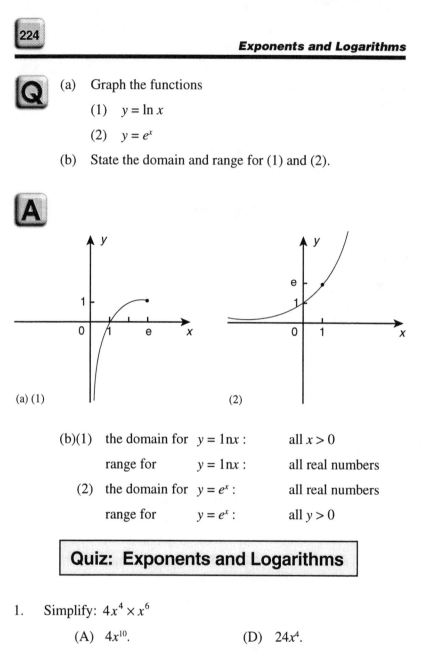

(a) (1)   (2)

(b)(1) the domain for $y = \ln x$:   all $x > 0$

range for $y = \ln x$:   all real numbers

(2) the domain for $y = e^x$:   all real numbers

range for $y = e^x$:   all $y > 0$

Quiz: Exponents and Logarithms

1. Simplify: $4x^4 \times x^6$

(A) $4x^{10}$.

(D) $24x^4$.

(B) $4x^{24}$.

(E) x^{22}.

(C) $16x^{10}$.

2. Simplify leaving no negative exponents: $\dfrac{x^{-2}y^3z^0}{2xy^{-4}z^{-1}}$

(A) $\dfrac{y^7z}{2x^3}$.

(D) $\dfrac{x^3y^7}{2z}$.

(B) 0.

(E) $\dfrac{y^7z^2}{2x^2}$.

(C) $\dfrac{xz}{2y^2}$.

3. Simplify leaving no negative exponents: $\left(\dfrac{x^{-1}y^2}{x^{-2}z^{-3}}\right)^{\frac{1}{5}}$

(A) $\sqrt{xy^2z^3}$.

(D) $x^5y^{10}z^{15}$.

(B) $\sqrt[5]{x^3y^2z^3}$.

(E) $x^6y^7z^8$.

(C) $\sqrt[5]{xy^2z^3}$.

4. Simplify leaving no negative exponents: $\left(\dfrac{4a^2b^{-3}}{9ab^4}\right)^{-\frac{1}{2}}$

(A) $\dfrac{3}{2}\sqrt{\dfrac{b}{a}}$.

(D) $\dfrac{3}{2}\sqrt[3]{ab}$.

(B) $\dfrac{2}{3b^3}\sqrt{\dfrac{a}{b}}$.

(E) $\dfrac{3}{2}b^3\sqrt{\dfrac{b}{a}}$.

(C) $\dfrac{2}{3}b^3\sqrt{\dfrac{a}{b}}$.

5. Let $f(x) = 1 + 2^{-x}$. What is the value of $f\left(\dfrac{1}{2}\right)$?

    (A) 0.

    (D) $\dfrac{1}{\sqrt{3}}$.

    (B) $\dfrac{1+\sqrt{2}}{\sqrt{2}}$.

    (E) $1+\sqrt{2}$.

    (C) $\sqrt{3}$.

6. If $m \neq 0$, then $(25)^{3m}(125)^{8m}(5)^m$ can be expressed as

    (A) 125^{12m}.

    (D) 25^{12m}.

    (B) 5^{20m}.

    (E) 5^{31m}.

    (C) 5^{19m}.

7. Evaluate the logarithmic expression $\dfrac{\log\left[(25)(15)\right]^2}{\log(25)+\log(15)}$.

    (A) 2.574.

    (D) 14.955.

    (B) 2.000.

    (E) Not solvable.

    (C) 9.375.

8. Solve for x: $\ln(\ln x) = 1$

    (A) 1.

    (D) e.

    (B) e^e.

    (E) No solution.

    (C) $\dfrac{1}{e}$.

9. One solution of the equation $27^{x^2+1} = 243$ is

   (A) $\sqrt{5}$.

   (D) $\dfrac{-5}{2}$.

   (B) $\dfrac{3}{2}$.

   (E) 5.

   (C) $-\sqrt{\dfrac{2}{3}}$.

10. What is the relationship between the two functions

   $f(x) = \log_2 x$ and $f(x) = 2^x$?

   (A) They are represented by the same graph.

   (B) Their graphs are perpendicular to each other.

   (C) They are inverses of each other.

   (D) They have the same domain and range.

   (E) There is no relationship.

ANSWER KEY

1.	(A)	6.	(E)
2.	(A)	7.	(B)
3.	(C)	8.	(B)
4.	(E)	9.	(C)
5.	(B)	10.	(C)

CHAPTER 8

Conic Sections

8.1 Introduction to Conic Sections

Conic sections are the curves formed when a plane intersects the surface of a right circular cone. As shown below, these curves are the circle, the ellipse, the parabola, and the hyperbola.

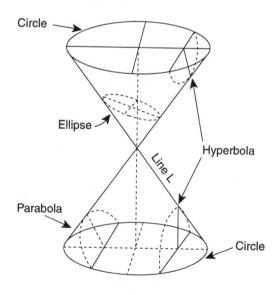

Figure 8.1

8.2 The Circle

A circle is defined to be the set of all points at a given distance from a given point. The given distance is called the radius, and the given point is called the center of the circle. Using the distance formula described in Section 2.3, it is easy to establish that an equation of a circle with center of (h,k) and radius of r is

$$(x-h)^2 + (y-k)^2 = r^2$$

An equation of this kind is said to be in standard form. Here are two examples which illustrate the practicality of the standard form.

EXAMPLE

Find an equation of the circle with center of $(2,-3)$ with radius of 6.

$$(x-2)^2 + [y-(-3)]^2 = 6^2$$
$$(x-2)^2 + (y+3)^2 = 36$$

Graph the equation listed below. (See Figure 8.2)

$$x^2 + y^2 - 6x + 10y + 30 = 0$$
$$(x^2 - 6x) + (y^2 + 10y) = -30$$

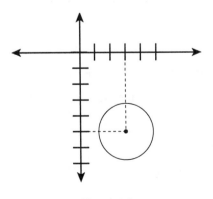

Figure 8.2

$$\left(x^2 - 6x + 9\right) + \left(y^2 + 10y + 25\right) = -30 + 34$$

$$\left(x - 3\right)^2 + \left(y + 5\right)^2 = 4$$

Notice, in the second example, how easy it is to graph

$$\left(x - 3\right)^2 + \left(y + 5\right)^2 = 4$$

and how difficult it would be to graph

$$x^2 + y^2 - 6x + 10y + 30 = 0$$

in that form.

Problem Solving Examples:

 Write equations of the following circles:

    (a) With center at (–1,3) and radius 9.

    (b) With center at (2,–3) and radius 5.

The equation of the circle with center at (a,b) and radius r is

$$\left(x - a\right)^2 + \left(y - b\right)^2 = r^2.$$

(a) Thus, the equation of the circle with center at (–1,3) and radius 9 is

$$\left[x - (-1)\right]^2 + \left(y - 3\right)^2 = 9^2$$

$$\left(x + 1\right)^2 + \left(y - 3\right)^2 = 81$$

(b) Similarly the equation of the circle with center at (2,–3) and radius 5 is

$$\left(x - 2\right)^2 + \left[y - (-3)\right]^2 = 5^2$$

$$\left(x - 2\right)^2 + \left(y + 3\right)^2 = 25$$

Q Find the center and radius of the circle

$$x^2 - 4x + y^2 + 8y - 5 = 0 \qquad (1)$$

A We can find the radius and the coordinates of the center by completing the square in both x and y. To complete the square in either variable, take half the coefficient of the variable term (i.e., the x term or the y term) and then square this value. The resulting number is then added to both sides of the equation. Completing the square in x:

$$\left[\frac{1}{2}(-4)\right]^2 = [-2]^2 = 4$$

Then equation (1) becomes:

$$\left(x^2 - 4x + 4\right) + y^2 + 8y - 5 = 0 + 4,$$

or

$$(x-2)^2 + y^2 + 8y - 5 = 4 \qquad (2)$$

Before completing the square in y, add 5 to both sides of equation (2):

$$(x-2)^2 + y^2 + 8y - 5 + 5 = 4 + 5$$
$$(x-2)^2 + y^2 + 8y = 9 \qquad (3)$$

Now, completing the square in y:

$$\left[\frac{1}{2}(8)\right]^2 = [4]^2 = 16$$

Then equation (3) becomes:

$$(x-2)^2 + \left(y^2 + 8y + 16\right) = 9 + 16,$$

or

$$(x-2)^2 + (y+4)^2 = 25 \qquad (4)$$

Note that the equation of a circle is:

$$(x-h)^2 + (y-k)^2 = r^2,$$

where (h,k) is the center of the circle and r is the radius of the circle. Equation (4) is in the form of the equation of a circle. Hence, equation (4) represents a circle with center $(2,-4)$ and radius = 5.

Find the equation of the circle of radius $r = 9$, with center on
$\ell : y = x$ and tangent to both coordinate axes.

 As seen in Figure 1, there are two such circles, 0 and 0'. Let
the centers of 0 and 0' be (a,b) and (c,d), respectively.

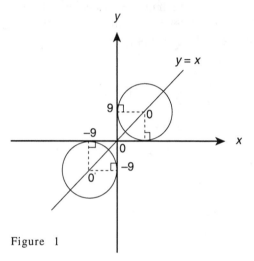

Figure 1

Since points (a,b) and (c,d) are on $\ell : y = x$, one has $a = b$ and $c = d$.

The equations will be

$$(x-a)^2 + (y-a)^2 = 9^2$$

and $(x-c)^2 + (y-c)^2 = 9^2$

Because the circles are tangent to both coordinate axes, one obtains $a = 9 = b$ and $c = -9 = d$. (See Figure 1.) Therefore, the equations are

$$(x-9)^2 + (y-9)^2 = 81 \text{ and}$$
$$(x+9)^2 + (y+9)^2 = 81$$

8.3 The Ellipse

An ellipse is defined to be the set of all points in a plane, the sum of whose distances from two fixed points is a constant. Each of the fixed points is called a focus. The plural of the word focus is foci. A simple method of constructing an ellipse comes directly from this definition. Mark two of the foci and call them F_1 and F_2, then insert a thumb tack at each focus. Next, take a string which is longer than the distance between F_1 and F_2 and tie one end at F_1 and the other end at F_2. Then, pull the string taut with a pencil and trace the ellipse. It will be oval shaped and will have two lines of symmetry and one point of symmetry.

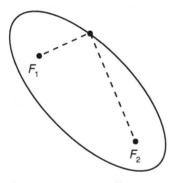

Figure 8.3

Using the distance formula, it is fairly easy to derive an equation of the ellipse with foci at $(c,0)$ and $(-c,0)$ with the sum of the distance of $2a$. The standard equation of such an ellipse is

$$\frac{x^2}{a^2} + \frac{y^2}{b^2} = 1,$$

where $b^2 = a^2 - c^2$. Obviously, $a^2 > b^2$. Here is a graph of such an ellipse.

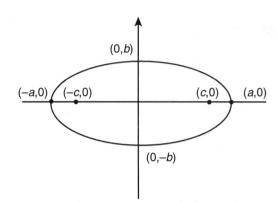

Figure 8.4

The center of the ellipse is the point midway between the foci. In this case, the center is at the origin. The segment from $(-a,0)$ to $(a,0)$ is called the major axis, and its length is $2a$. The segment from $(0,-b)$ to $(0,b)$ is called the minor axis, and its length is $2b$. In the ellipse, the x-axis and the y-axis are lines of symmetry, and the origin is a point of symmetry. The eccentricity is defined to be $\dfrac{c}{a}$. When this ratio is close to 0, the ellipse resembles a circle, but when this ratio is close to 1, the ellipse is elongated.

It is very easy to graph an ellipse when the standard equation is given. In most cases, it is desirable to convert to that form. This process is illustrated in the example below.

EXAMPLE

Graph

$$9x^2 + 16y^2 = 144$$

$$\frac{9x^2}{144} + \frac{16y^2}{144} = \frac{144}{144}$$

$$\frac{x^2}{16} + \frac{y^2}{9} = 1$$

$$\frac{x^2}{4^2} + \frac{y^2}{3^2} = 1$$

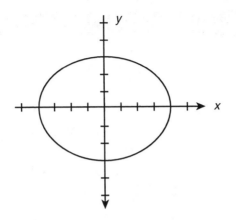

Figure 8.5

For an ellipse with foci at $(0,c)$ and $(0,-c)$, and with the sum of distances as $2a$, the standard equation is

$$\frac{y^2}{a^2} + \frac{x^2}{b^2} = 1$$

Here is a table which illustrates what happens when the center of the ellipse is not at the origin.

Center of ellipse	Sum of distance	Foci	Standard equation
(h,k)	$2a$	$(h+c,k)$ and $(h-c,k)$	$\dfrac{(x-h)^2}{a^2} + \dfrac{(y-k)^2}{b^2} = 1$
(h,k)	$2a$	$(h,k+c)$ and $(h,k-c)$	$\dfrac{(y-k)^2}{a^2} + \dfrac{(x-h)^2}{b^2} = 1$

In both cases, $b^2 = a^2 - c^2$.

Problem Solving Examples:

 Discuss the graph of $\dfrac{x^2}{25} + \dfrac{y^2}{9} = 1$.

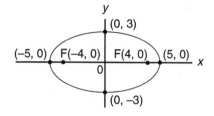

 Since this is an equation of the form $\dfrac{x^2}{a^2} + \dfrac{y^2}{b^2} = 1$, with $a = 5$ and $b = 3$, it represents an ellipse. The simplest way to sketch the curve is to find its intercepts.

If we set $x = 0$, then

$$y = \sqrt{\left(1 - \frac{x^2}{25}\right)9} = \sqrt{\left(1 - \frac{0^2}{25}\right)9} = \pm 3$$

so that the y-intercepts are at $(0,3)$ and $(0,-3)$. Similarly the x-intercepts are found for $y = 0$:

$$x = \sqrt{\left(1 - \frac{y^2}{9}\right)25}$$

$$= \sqrt{\left(1 - \frac{0^2}{9}\right)25}$$

$$= \pm 5$$

to be at $(5,0)$ and $(-5,0)$ (see figure). To locate the foci we note that

$$c^2 = a^2 - b^2 = 5^2 - 3^2$$
$$c^2 = 25 - 9 = 16$$
$$c = \pm 4.$$

The foci lie on the major axis of the ellipse. In this case it is the x-axis since $a = 5$ is greater than $b = 3$. Therefore, the foci are $(\pm c, 0)$, that is, at $(-4, 0)$ and $(4, 0)$. The sum of the distances from any point on the curve to the foci is $2a = 2(5) = 10$.

 In the equation of an ellipse,
$$4x^2 + 9y^2 - 16x + 18y - 11 = 0,$$

determine the standard form of the equation, and find the values of a, b, c, and e.

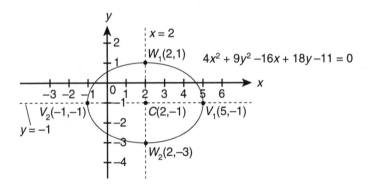

A By completing the squares, we can arrive at the standard form of the equation, from which the values of the parameters can be determined. Thus,

$$4\left(x^2 - 4x + 4\right) + 9\left(y^2 + 2y + 1\right) = 36.$$

or

$$4(x-2)^2 + 9(y+1)^2 = 36. \text{ Dividing by } 36,$$

$$\frac{(x-2)^2}{9} + \frac{(y+1)^2}{4} = 1.$$

Thus, the center of the ellipse is at (2,–1). Comparing this equation with the general form,

$$\frac{x^2}{a^2} + \frac{y^2}{b^2} = 1, \text{ where } a > b, \text{ we see that } a = 3, b = 2.$$

$$c = \sqrt{a^2 - b^2} = \sqrt{5}.$$

Finally, $e = \dfrac{c}{a} = \dfrac{\sqrt{5}}{3} \approx 0.745.$

 Find the equation of the ellipse which has vertices V_1 (–2,6), V_2 (–2,–4), and foci F_1 (–2,4), F_2 (–2,–2). (See figure.)

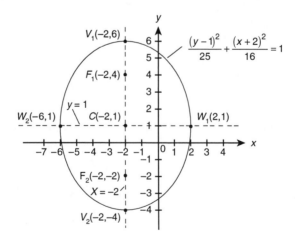

A The major axis is on the line $x = -2$, and the center is at $(-2,1)$. Hence, a, the length of the semimajor axis, equals the difference between the y-coordinates of V_1 (say) and the center, i.e., $a = 5$. From the coordinates of the foci, $c = 3$. Since $b^2 = a^2 - c^2$, $b = \sqrt{25-9} = 4$, and the ends of the minor axis, on $y = 1$, are at W_1 $(2,1)$ and W_2 $(-6,1)$. The equation can now be written, in the form

$$\frac{(y-k)^2}{a^2} + \frac{(x-h)^2}{b^2} = 1$$

$$\frac{(y-1)^2}{25} + \frac{(x+2)^2}{16} = 1.$$

8.4 The Parabola

A parabola is defined to be the set of all points in a plane equally distant from a fixed line and a fixed point not on the line. The line is called the directrix and the point is called the focus. The axis of a parabola is the line which passes through the focus and which is perpendicular to the directrix. The vertex of the parabola is at the point which is halfway between the focus and the directrix. A picture of a parabola is shown in Figure 8.6.

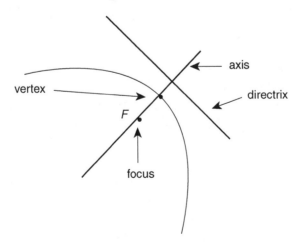

Figure 8.6

It is somewhat common for the undirected distance between the directrix and the focus to be labelled $2c$. (this means that $c > 0$.) Here is a table which relates the standard equation to the position of the vertex, the position of the focus and the equation of the directrix.

Directrix	Focus	Vertex	Standard Equation	Direction Parabola Opens
$y = -c$	$(0, c)$	$(0, 0)$	$x^2 = 4cy$	Up
$x = -c$	$(c, 0)$	$(0, 0)$	$y^2 = 4cx$	Right
$y = c$	$(0, -c)$	$(0, 0)$	$x^2 = -4cx$	Down
$x = c$	$(-c, 0)$	$(0, 0)$	$y^2 = -4cx$	Left
$y = -c + k$	$(h, c + k)$	(h, k)	$(x - h)^2 = 4c(y - k)$	Up
$x = -c + h$	$(c + h, k)$	(h, k)	$(y - k)^2 = 4c(x - h)$	Right
$y = c + k$	$(h, -c + k)$	(h, k)	$(x - h)^2 = -4c(y - k)$	Down
$x = c + h$	$(-c + h, k)$	(h, k)	$(y - k)^2 = -4c(x - h)$	Left

The table above illustrates that once an equation for a parabola is in standard form, it is easy to characterize the parabola. Consider the following.

$$y^2 - 6y - 6x + 39 = 0$$
$$y^2 - 6y = 6x - 39$$
$$y^2 - 6y + 9 = 6x - 30$$
$$(y - 3)^2 = 6(x - 5)$$

This is a parabola with vertex at (5,3) and focus at $(6\frac{1}{2}, 3)$. The equation of the directrix is $x = 3\frac{1}{2}$, and the parabola opens to the right.

In general, equations of the form

$$y^2 + Dx + Ey + F = 0$$

are parabolas which open to the right or left, and equations of the form

$$x^2 + Dx + Ey + F = 0$$

are parabolas which open up or down.

Problem Solving Examples:

Q Construct the graph of the function defined by

$$y = x^2 - 6x + 10.$$

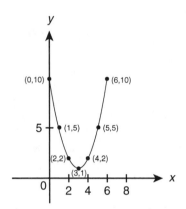

We are given the function $y = x^2 - 6x + 10$.

The most general form of the quadratic function is $y = ax^2 + bx + c$ where a, b, and c are constants. If a is positive, the curve opens upward and it is U-shaped. If a is negative, the curve opens downward and it is inverted U-shaped.

Since $a = 1 > 0$ in the given equation, the graph is a parabola that opens upward. To determine the pairs of values (x,y) which satisfy this equation, we express the quadratic function in terms of the square of a linear function of x.

$$y = x^2 - 6x + 10 = x^2 - 6x + 9 + 1$$
$$= (x - 3)^2 + 1$$

y is least when $x - 3 = 0$ This is true because the square of any number, be it positive or negative, is a positive number. Therefore y would always be greater than or equal to 1. Thus the minimum value of y is one when $x - 3 = 0$ or $x = 3$.

In order to plot the curve, we select values for x and calculate the corresponding y values. (See the table.)

x	$x^2 - 6x + 10 =$	y
0	$(0)^2 - 6(0) + 10$	10
1	$(1)^2 - 6(1) + 10$	5
2	$(2)^2 - 6(2) + 10$	2
3	$(3)^2 - 6(3) + 10$	1
4	$(4)^2 - 6(4) + 10$	2
5	$(5)^2 - 6(5) + 10$	5
6	$(6)^2 - 6(6) + 10$	10

The points and graphs determined by the table are shown in the accompanying figure.

Q Find the coordinates of the maximum point of the curve $y = -3x^2 - 12x + 5$, and locate the axis of symmetry.

A The curve is defined by a second degree equation. The coefficient of the x^2 term is negative. Hence, the graph of this curve is a parabola opening downward. The maximum point of the curve occurs at the vertex and has the x-coordinate:

$$-\frac{\text{coefficient of } x \text{ term}}{2(\text{coefficient of } x^2 \text{ term})} = -\frac{b}{2a} = -\frac{-12}{2(-3)} = \frac{12}{-6} = -2.$$

For $x = -2$, $y = -3(-2)^2 - 12(-2) + 5 = 17$. Hence the coordinates of the vertex are $(-2, 17)$. The curve is symmetric with respect to the vertical line through its vertex. The axis of symmetry of this curve is the vertical line through the point $(-2, 17)$, i.e., the line $x = -2$.

Discuss the graph of the equation $y^2 = 12x$.

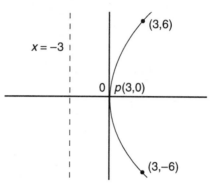

 The equation written as $x = \dfrac{1}{12}y^2$ is a quadratic equation with the coefficient of the y^2 term positive. Therefore the graph is a parabola opening to the right. Since $f(x) = -f(x)$ the parabola is symmetric with respect to the x-axis. Point $(0,0)$ satisfies the equation and lies on the axis of symmetry. Hence the vertex of the parabola is at $(0,0)$ (see figure). The focus of the parabola lies on the axis of symmetry, $y = 0$, at the point $(p,0)$ where $4p$ = coefficient of x in the original equation: $4p = 12$, $p = 3$. Therefore the focus is at $(3,0)$. The directrix is the vertical line $x = -p = -3$. When $x = 3$, $y = \pm\sqrt{12x} = \pm\sqrt{12(3)} = \pm 6$. Therefore the points $(3,6)$ and $(3,-6)$ are points on the graph. The graph of this parabola is not the graph of a function, since for any given value of x there is more than one corresponding value of y.

 Write the equation of the parabola whose focus has coordinates $(0,2)$ and whose directrix has equation $y = -2$. (See figure.)

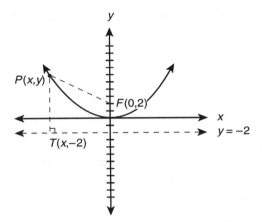

A Since, by definition, each point lying on a parabola is equidistant from both the focus and directrix of the parabola, the origin must lie on the specific parabola described in the statement of the problem (see figure).

To find the equation of the parabola, choose a point $P(x,y)$ lying on the parabola (see figure). By definition, then, the distance PT must equal the distance PF, where T lies on the directrix, directly below P. Since T also lies on $y = -2$, it has coordinates $(x,-2)$. Using the distance formula, we find $PF = PT$

$$\sqrt{x^2 + (y-2)^2} = \sqrt{(y+2)^2}.$$

Squaring both sides of this equation, we obtain

$$x^2 + (y-2)^2 = (y+2)^2.$$

Expanding this, we get

$$x^2 + y^2 - 4y + 4 = y^2 + 4y + 4.$$

Subtracting y^2, $-4y$, and 4 from each side of this equation yields

$$x^2 = 8y.$$

Dividing both sides of this equation by 8 gives the equation of the parabola,

$$y = \frac{1}{8}x^2.$$

8.5   The Hyperbola

The hyperbola is defined to be the set of all points in a plane, the difference of whose distances from two fixed points is a constant. If the foci are at $(c,0)$ and $(-c,0)$, and if the difference of the distances is $2a$ with $0 < a < c$, then the standard equation for this hyperbola is

$$\frac{x^2}{a^2} - \frac{y^2}{b^2} = 1,$$

where $b^2 = c^2 - a^2$. A graph of the hyperbola follows.

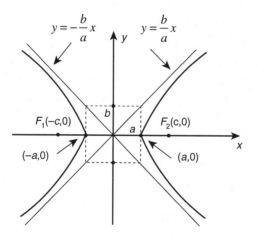

Figure 8.7

Notice that the hyperbola has two branches.

For the hyperbola, the focal axis is the line through the foci, the center is the point midway between the foci, and the conjugate axis is the line through the center which is perpendicular to the focal axis. The hyperbola has symmetry with respect to those lines and that point. For the hyperbola pictured above, the focal axis is the x-axis, the conjugate axis is the y-axis, and the center is at the origin. Asymptotes are described in Section 2.5. Notice that, in this case, the lines

$$y = \frac{b}{a}x \quad \text{and} \quad y = \frac{-b}{a}x$$

are asymptotes for the hyperbola pictured.

If the foci are at $(0,c)$ and $(0,-c)$, and if the difference between the distances is still $2a$ with $0 < a < c$, then the standard equation for the hyperbola is

$$\frac{y^2}{a^2} - \frac{x^2}{b^2} = 1,$$

where $b^2 = c^2 - a^2$.

In this case, the center is at the origin, the focal axis is the y-axis, the conjugate axis is the x-axis, and the hyperbola has the lines

$$y = \frac{a}{b}x \quad \text{and} \quad y = \frac{-a}{b}x$$

as asymptotes. A graph of this hyperbola follows.

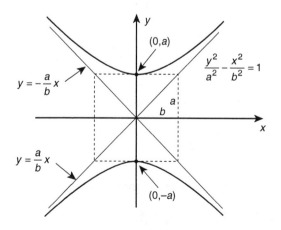

Figure 8.8

The standard form of the equation of the hyperbola with center at (h,k) and with focal axis parallel to the x-axis is

$$\frac{(x-h)^2}{a^2} - \frac{(y-k)^2}{b^2} = 1,$$

and the standard form for the equation of the hyperbola with center at (h,k) and with focal axis parallel to the y-axis is

$$\frac{(y-k)^2}{a^2} - \frac{(x-h)^2}{b^2} = 1.$$

Problem Solving Examples:

 Consider the equation
$$x^2 - 4y^2 + 4x + 8y + 4 = 0.$$

Express this equation in standard form, and determine the center, the vertices, the foci, and the eccentricity of this hyperbola. Describe the fundamental rectangle and find the equations of the 2 asymptotes.

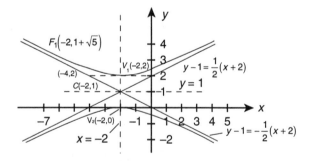

 Rewrite the equation by completing the squares, i.e.,

$$\left(x^2 + 4x + 4\right) - 4\left(y^2 - 2y + 1\right) = -4$$

or $\qquad (x+2)^2 - 4(y-1)^2 = -4$

or, dividing, rearranging terms,

$$\frac{(y-1)^2}{1} - \frac{(x+2)^2}{4} = 1.$$

The center, located at (h, k) in the equation

$$\frac{(y-k)^2}{a^2} - \frac{(x-h)^2}{b^2} = 1$$ is, therefore, at $(-2,1)$. Furthermore, $a = 1$, $b = 2$. Thus,

$$c = \sqrt{1^2 + 2^2} = \sqrt{5}, \text{ and } e = \sqrt{5}.$$

The vertices are displaced $\pm a$ from the center, while the foci are displaced $\pm c$ (along the transverse -or focal-axis). Therefore, the vertices are $(-2, 1 \pm 1)$ and the foci are $(-2, 1 \pm \sqrt{5})$.

By definition, the fundamental rectangle is the rectangle whose vertices are at $(h \pm b, k \pm a)$. Hence, in this example, the coordinates of the vertices of the rectangle are $(0,2)$, $(-4, 2)$, $(-4, 0)$, and $(0,0)$. The equations of the two asymptotes are determined by finding the slopes of the lines passing through the center of the hyperbola and two of the vertices of its fundamental rectangle (see figure). Then,

$$m = \frac{\Delta y}{\Delta x} = \pm \frac{1}{2}$$

gives the two slopes and the point-slope form, choosing the point $(-2,1)$ which is common to both asymptotes, gives

$$y - 1 = \pm \frac{1}{2}(x + 2).$$

 Find the equation of the hyperbola with vertices $V_1(8,0)$, $V_2(2,0)$ and eccentricity $e = 2$.

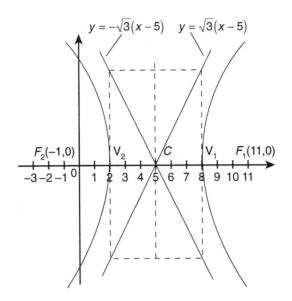

 There are two basic forms of the equation of an hyperbola that is not rotated with respect to the coordinate axes:

$$\frac{(x-h)^2}{a^2} - \frac{(y-k)^2}{b^2} = 1 \text{ and } \frac{(y-k)^2}{a^2} - \frac{(x-h)^2}{b^2} = 1.$$

Which form is appropriate depends upon whether the transverse axis is parallel to the x-axis or to the y-axis, respectively. To determine the equation of an hyperbola, then, it is necessary to first discover which equation applies, then to solve for the constants h, k, a, c. The information about the vertices implies that the transverse axis is the x-axis.

Thus, the first form of the equation for an hyperbola applies. In this case, the center, which is the average of the vertices is at (5,0). The distance between the vertices is $2a = 6$; therefore, $a = 3$. In order to determine the value of b, we use the relation between eccentricity, e, c, and a: $e = c/a$. Thus, $c = e \times a = 2 \times 3 = 6$.

$$b = \sqrt{c^2 - a^2} = \sqrt{6^2 - 3^2} = \sqrt{27} \approx 5.2$$

Substituting for h, k, a, b, we have

$$\frac{(x-5)^2}{9} - \frac{y^2}{27} = 1,$$

or $3x^2 - y^2 - 30x + 48 = 0$.

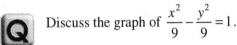 Discuss the graph of $\dfrac{x^2}{9} - \dfrac{y^2}{9} = 1$.

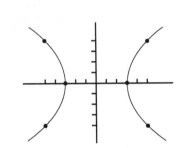

A $\dfrac{x^2}{9} - \dfrac{y^2}{9} = 1$ is an equation of the form $\dfrac{x^2}{a^2} - \dfrac{y^2}{b^2} = 1$ with $a = 3$ and $b = 3$. Therefore the graph is a hyperbola. The x-intercepts are found by setting $y = 0$:

$$\frac{x^2}{9} - \frac{y^2}{9} = 1$$
$$x^2 = 9$$
$$x = \pm 3.$$

Thus, the x-intercepts are at $(-3,0)$ and $(3,0)$. There are no y-intercepts since for $x = 0$ there are no real values of y satisfying the equation, i.e., no real value of y satisfies

$$\frac{0^2}{9} - \frac{y^2}{9} = 1$$

$$y^2 = -9, \quad y = \sqrt{-9}.$$

Solving the original equation for y:

$$y = \sqrt{\left(1 - \frac{x^2}{9}\right)(-9)} \quad \text{or} \quad y = \sqrt{x^2 - 9}$$

shows that there will be no permissible values of x in the interval $-3 < x < 3$. Such values of x do not yield real values for y. For $x = 5$ and $x = -5$ use the equation for y to obtain the ordered pairs (5,4), (5,-4), (-5,4), and (-5,-4) as indicated in the figure. The foci of the hyperbola are located at $(\pm c, 0)$, where

$$c^2 = a^2 + b^2$$

$$c^2 = 3^2 + 3^2 = 9 + 9 = 18$$

$$c = \pm\sqrt{18} = \pm 3\sqrt{2}.$$

Therefore, the foci are at $\left(-3\sqrt{2}, 0\right)$ and $\left(3\sqrt{2}, 0\right)$.

 Draw the graph of $xy = 6$.

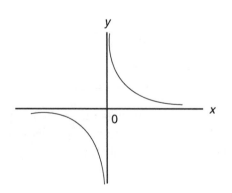

A Since the product is positive the values of x and y must have the same sign, that is, when x is positive y must also be positive and when x is negative then y is also negative. Moreover, neither x nor y can be zero (or their product would be zero not 6), so that the graph never touches the coordinate axes. Solve for y and we obtain $y = \dfrac{6}{x}$. Substituting values of x into this equation we construct the following chart:

x:	-6	-3	-2	-1	1	2	3	6
y:	-1	-2	-3	-6	6	3	2	1

The graph is obtained by plotting the above points and then joining them with a smooth curve, remembering that the curve can never cross a coordinate axis. The graph of the equation, $xy = k$, is a hyperbola for all nonzero real values of k. If k is negative, then x and y must have opposite signs, and the graph is in the second and fourth quadrants as opposed to the first and third.

Quiz: Conic Sections

1.   What is the center and radius of the circle given by
$x^2 + y^2 + 6x - 10y = -32$?

(A)   center $(-3,5)$ radius: $\sqrt{2}$

(B)   center $(3,-5)$ radius: $\sqrt{2}$

(C)   center $(-3,5)$ radius: 2

(D)   center $(3,5)$ radius: 2

(E)   center $(5,3)$ radius: $\sqrt{2}$

2. A circle is centered at point (3,4). The endpoints of a diameter are A and B. The coordinates of Point A are $(-2,1)$. What are the coordinates of point B?

   (A) (6,3).   (D) (9,3).

   (B) (8,7).   (E) (5,8).

   (C) (4,5).

3. An equation of a circle with center $\left(-\dfrac{2}{3}, \dfrac{1}{2}\right)$ and radius $\sqrt{7}$ is:

   (A) $3x^2 + 4x + 3y^2 = 3y = 227$

   (B) $36x^2 - 48x + 36y^2 + 36y = 259$

   (C) $36x^2 + 48x + 36y^2 - 36y = 227$

   (D) $3x^2 - 2x + y^2 + 2y = 7$

   (E) $48x^2 - 36x + 36y^2 + 36y = 227$

4. The equation $6x^2 - 24x + 4y^2 + 32y = -52$ represents which of the following?

   (A) circle.   (D) ellipse.

   (B) hyperbola.   (E) none of these.

   (C) parabola.

5. The equation $x^2 - 2x + 2y^2 + 8y = -5$ is the equation of an ellipse. What are the coordinates of the foci of the ellipse?

    (A) $(3,-2)$ and $(-1,-2)$

    (B) $\left(1+\sqrt{2},-2\right)$ and $\left(1-\sqrt{2},-2\right)$

    (C) $\left(1-\sqrt{3},2\right)$ and $\left(1,-2-\sqrt{2}\right)$

    (D) $\left(1,-\sqrt{3},2\right)$ and $\left(1+\sqrt{3},2\right)$

    (E) $\left(-1+\sqrt{2},2\right)$ and $\left(-1-\sqrt{2},2\right)$

6. What is the equation of a parabola which is symmetrical about the y-axis and passes through points $(0,-2)$ and $(2,0)$?

    (A) $y = x^2 - 2$ (D) $y = 2x^2 - 1$

    (B) $y = \dfrac{1}{2}x^2 - 2$ (E) $y = 2x^2 - \dfrac{1}{2}$

    (C) $y = \dfrac{3}{2}x^2 + \dfrac{1}{2}$

7. Find the center of the ellipse given by the equation

    $$x^2 + x + 3y + 2y^2 - 1 = 0.$$

    (A) $\left(\dfrac{1}{2}, -\dfrac{3}{2}\right)$ (D) $\left(-\dfrac{1}{2}, -\dfrac{3}{4}\right)$

    (B) $\left(\dfrac{1}{4}, \dfrac{3}{2}\right)$ (E) $\left(1, -\dfrac{3}{4}\right)$

    (C) $\left(\dfrac{1}{2}, -\dfrac{3}{4}\right)$

8. A parabola has the equation $y = \dfrac{-1}{2}x^2 + 3x - \dfrac{13}{2}$. What is the directrix of this parabola?

    (A) $y = \dfrac{-5}{2}$

    (B) $x = \dfrac{-3}{2}$

    (C) $x = \dfrac{-5}{2}$

    (D) $y = \dfrac{-3}{2}$

    (E) $y = \dfrac{5}{2}$

9. What are the equations of the asymptotes for the hyperbola given by $\dfrac{(x+1)^2}{9} - \dfrac{(y-2)^2}{4} = 1$?

    (A) $y = \dfrac{1}{3}x + \dfrac{2}{3}$ and $y = \dfrac{-1}{3}x - \dfrac{1}{3}$

    (B) $y = \dfrac{2}{3}x + \dfrac{8}{3}$ and $y = \dfrac{-2}{3}x + \dfrac{4}{3}$

    (C) $y = \dfrac{8}{3}x + \dfrac{-2}{3}$ and $y = \dfrac{-8}{3}x + \dfrac{4}{3}$

    (D) $y = \dfrac{3}{3}x + 3$ and $y = \dfrac{-3}{2}x + \dfrac{1}{2}$

    (E) none of these

10. Find a, b for the parabola

$$y = ax^2 + bx + 3$$

if the vertex is (2,4).

(A) $a = -\dfrac{1}{4}, b = 2$

(D) $a = -\dfrac{1}{3}, b = 1$

(B) $a = -1, b = -2$

(E) $a = -\dfrac{1}{4}, b = 1$

(C) $a = 1, b = 2$

ANSWER KEY

1.	(A)	6.	(B)
2.	(B)	7.	(D)
3.	(C)	8.	(D)
4.	(D)	9.	(B)
5.	(B)	10.	(E)

CHAPTER 9

Matrices and
Determinants

9.1 Matrices and Matrix Operations

A matrix is a rectangular array of numbers, usually real numbers. These numbers are called entries. Matrices are classified according to their number of rows and columns. An m by n matrix has m rows and n columns, and such a matrix is said to have size m by n. Two matrices of the same size are equal when the corresponding entries are equal.

To add matrices of the same size, merely add the corresponding entries. The following is a description of properties concerning the addition of matrices.

Suppose A, B, and C are arbitrary m by n matrices, **0** is the m by n matrix with 0 as each entry, and $-A$ is the m by n matrix in which each entry is the opposite, or negative, of the corresponding entry in A. Then

$$A + B = B + A$$
$$(A + B) + C = A + (B + C)$$
$$A + O = O + A = A$$
$$A + (-A) = -A + A = O$$

To subtract two matrices of the same size, merely subtract the corresponding entries. To multiply a matrix by a number, multiply each entry in the matrix by that number. This is called scalar multiplication, and the number (usually real number) is called a scalar. The following is a significant scalar multiplication property.

If A is an m by n matrix and a and b are scalars, then

$$(ab)A = a(bA)$$

The standard product of two matrices is much more difficult to find than the sum or difference of two matrices, or the scalar product of a number and a matrix. If A is an m by p matrix and B is a p by n matrix, then AB is an m by n matrix where the entry in the ith row and jth column of AB is the sum of all the products found from multiplying entries in the ith row of A with the jth column of B in order. Note, for the product AB to exist, A must have the same number of columns as B has rows. Of course, this means that when A and B are of the same size, the product may not even exist. For that reason, it is somewhat common to primarily confine matrix multiplication to square matrices. Here are two properties concerning products of square matrices.

Suppose A, B, and C are arbitrary n by n matrices and I is the n by n matrix with entries in the first row first column, second row second column, third row third column, ... n^{th} row n^{th} column all 1, and with all other entries 0. Then

$$(AB)C = A(BC)$$

$$AI = IA = A$$

It is somewhat obvious that there are n by n matrices A and B, such that $AB \neq BA$. In many cases, there exists an n by n matrix A^{-1} with the property that $AA^{-1} = A^{-1}A = I$. Section 9.3 is devoted to determining when square matrices have multiplicative inverses and how to find them when they do exist.

Problem Solving Examples:

If $A = \begin{bmatrix} 2 & -2 & 4 \\ -1 & 1 & 1 \end{bmatrix}$ and $B = \begin{bmatrix} 0 & 1 & -3 \\ 1 & 3 & 1 \end{bmatrix}$, find $2A + B$.

For an $m \times n$ matrix, $A = (\alpha_{ij})$, we know $cA = (c\alpha_{ij})$. Hence,

$$2A = 2\begin{bmatrix} 2 & -2 & 4 \\ -1 & 1 & 1 \end{bmatrix} = \begin{bmatrix} 2 \times 2 & 2 \times (-2) & 2 \times 4 \\ 2 \times (-1) & 2 \times 1 & 2 \times 1 \end{bmatrix} = \begin{bmatrix} 4 & -4 & 8 \\ -2 & 2 & 2 \end{bmatrix}.$$

For two $m \times n$ matrices, $A = (\alpha_{ij})$ and $B = (\beta_{ij})$, the ith row of the matrix $A + B$ is given by $(\alpha_{i1} + \beta_{i1}, \ldots, \alpha_{i1} + \beta_{in})$.

Thus,

$$2A + B = \begin{bmatrix} 4 & -4 & 8 \\ -2 & 2 & 2 \end{bmatrix} + \begin{bmatrix} 0 & 1 & -3 \\ 1 & 3 & 1 \end{bmatrix} = \begin{bmatrix} 4+0 & -4+1 & 8-3 \\ -2+1 & 2+3 & 2+1 \end{bmatrix}$$

$$2A + B = \begin{bmatrix} 4 & -3 & 5 \\ -1 & 5 & 3 \end{bmatrix}$$

Show that

(a) $A + B = B + A$ where

$$A = \begin{bmatrix} 3 & 1 & 1 \\ 2 & -1 & 1 \end{bmatrix}; \quad B = \begin{bmatrix} 4 & 2 & -1 \\ 0 & 0 & 2 \end{bmatrix}.$$

(b) $(A + B) + C = A + (B + C)$ where

$$A = \begin{bmatrix} -2 & 6 \\ 2 & 1 \end{bmatrix}, \quad B = \begin{bmatrix} 2 & 1 \\ 0 & 3 \end{bmatrix} \text{ and } C = \begin{bmatrix} -1 & 0 \\ 7 & 2 \end{bmatrix}.$$

(c) If A and the zero matrix $(0ij)$ have the same size, then $A + 0 = A$ where

$$A = \begin{bmatrix} 2 & 1 \\ 1 & 2 \end{bmatrix}.$$

(d) $A + (-A) = 0$ where

$$A = \begin{bmatrix} 2 & 1 \\ 1 & 2 \end{bmatrix}.$$

(e) $(ab)A = a(bA)$ where $a = -5$, $b = 3$ and

$$A = \begin{bmatrix} 6 & -1 & 0 \\ 1 & 2 & 1 \end{bmatrix}.$$

(f) Find B if $2A - 3B + C = 0$ where

$$A = \begin{bmatrix} -1 & 3 \\ 0 & 0 \end{bmatrix} \text{ and } C = \begin{bmatrix} -2 & -1 \\ -1 & 1 \end{bmatrix}.$$

 (a) By the definition of matrix addition,

$$A + B = \begin{bmatrix} 3 & 1 & 1 \\ 2 & -1 & 1 \end{bmatrix} + \begin{bmatrix} 4 & 2 & -1 \\ 0 & 0 & 2 \end{bmatrix}$$

$$= \begin{bmatrix} 3+4 & 1+2 & 1+(-1) \\ 2+0 & -1+0 & 1+2 \end{bmatrix}$$

$$= \begin{bmatrix} 7 & 3 & 0 \\ 2 & -1 & 3 \end{bmatrix}$$

and

$$B + A = \begin{bmatrix} 4 & 2 & -1 \\ 0 & 0 & 2 \end{bmatrix} + \begin{bmatrix} 3 & 1 & 1 \\ 2 & -1 & 1 \end{bmatrix}$$

$$= \begin{bmatrix} 4+3 & 2+1 & -1+1 \\ 0+2 & 0+(-1) & 2+1 \end{bmatrix} = \begin{bmatrix} 7 & 3 & 0 \\ 2 & -1 & 3 \end{bmatrix}$$

Thus, $A + B = B + A$.

(b)

$$A + B = \begin{bmatrix} -2 & 6 \\ 2 & 1 \end{bmatrix} + \begin{bmatrix} 2 & 1 \\ 0 & 3 \end{bmatrix}$$

$$= \begin{bmatrix} -2+2 & 6+1 \\ 2+0 & 1+3 \end{bmatrix} = \begin{bmatrix} 0 & 7 \\ 2 & 4 \end{bmatrix}$$

and

$$(A + B) + C = \begin{bmatrix} 0 & 7 \\ 2 & 4 \end{bmatrix} + \begin{bmatrix} -1 & 0 \\ 7 & 2 \end{bmatrix} = \begin{bmatrix} 0+(-1) & 7+0 \\ 2+7 & 4+2 \end{bmatrix} = \begin{bmatrix} -1 & 7 \\ 9 & 6 \end{bmatrix}.$$

$$B + C = \begin{bmatrix} 2 & 1 \\ 0 & 3 \end{bmatrix} + \begin{bmatrix} -1 & 0 \\ 7 & 2 \end{bmatrix} = \begin{bmatrix} 2+(-1) & 1+0 \\ 0+7 & 3+2 \end{bmatrix} = \begin{bmatrix} 1 & 1 \\ 7 & 5 \end{bmatrix}$$

and

$$A + (B + C) = \begin{bmatrix} -2 & 6 \\ 2 & 1 \end{bmatrix} + \begin{bmatrix} 1 & 1 \\ 7 & 5 \end{bmatrix} = \begin{bmatrix} -2+1 & 6+1 \\ 2+7 & 1+5 \end{bmatrix} = \begin{bmatrix} -1 & 7 \\ 9 & 6 \end{bmatrix}.$$

Thus, $(A + B) + C = A + (B + C)$.

(c) An $m \times n$ matrix all of whose elements are zeros is called a zero matrix and is usually denoted by $_m\mathbf{0}_n$.

$$A = \begin{bmatrix} 2 & 1 \\ 1 & 2 \end{bmatrix} \qquad 0 = \begin{bmatrix} 0 & 0 \\ 0 & 0 \end{bmatrix}.$$

Thus,

$$A + 0 = \begin{bmatrix} 2 & 1 \\ 1 & 2 \end{bmatrix} + \begin{bmatrix} 0 & 0 \\ 0 & 0 \end{bmatrix} = \begin{bmatrix} 2+0 & 1+0 \\ 1+0 & 2+0 \end{bmatrix} = \begin{bmatrix} 2 & 1 \\ 1 & 2 \end{bmatrix}.$$

Hence, $A + 0 = A$.

(d) $-A = -1 \times \begin{bmatrix} 2 & 1 \\ 1 & 2 \end{bmatrix} = \begin{bmatrix} -1 \times 2 & -1 \times 1 \\ -1 \times 1 & -1 \times 2 \end{bmatrix} = \begin{bmatrix} -2 & -1 \\ -1 & -2 \end{bmatrix}.$

Thus,

$$A + (-A) = \begin{bmatrix} 2 & 1 \\ 1 & 2 \end{bmatrix} + \begin{bmatrix} -2 & -1 \\ -1 & -2 \end{bmatrix} = \begin{bmatrix} 2+(-2) & 1+(-1) \\ 1+(-1) & 2+(-2) \end{bmatrix} = \begin{bmatrix} 0 & 0 \\ 0 & 0 \end{bmatrix}$$

Therefore, $A + (-A) = 0$.

(e) If $A = \begin{bmatrix} a_1 & b_1 \\ c_1 & d_1 \end{bmatrix}$ and a is any scalar from a field, aA is defined by

$$aA = \begin{bmatrix} aa_1 & ab_1 \\ ac_1 & ad_1 \end{bmatrix}.$$

So

$$bA = 3 \begin{bmatrix} 6 & -1 & 0 \\ 1 & 2 & 1 \end{bmatrix} = \begin{bmatrix} 3 \times 6 & 3 \times (-1) & 3 \times 0 \\ 3 \times 1 & 3 \times 2 & 3 \times 1 \end{bmatrix}$$
$$= \begin{bmatrix} 18 & -3 & 0 \\ 3 & 6 & 3 \end{bmatrix}$$

and

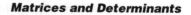

$$a(bA) = -5\begin{bmatrix} 18 & -3 & 0 \\ 3 & 6 & 3 \end{bmatrix} = \begin{bmatrix} -90 & 15 & 0 \\ -15 & -30 & -15 \end{bmatrix}$$

$$(ab)A = ((-5)(3))\begin{bmatrix} 6 & -1 & 0 \\ 1 & 2 & 1 \end{bmatrix} = -15\begin{bmatrix} 6 & -1 & 0 \\ 1 & 2 & 1 \end{bmatrix} = \begin{bmatrix} -90 & 15 & 0 \\ -15 & -30 & -15 \end{bmatrix}$$

Thus, $(ab)A = a(bA)$.

(f) $2A - 3B + C = 2A + C - 3B = 0$ since matrix addition is commutative. Now, add $3B$ to both sides of the equation,

$$2A + C - 3B = 0,$$

to obtain $2A + C - 3B + 3B = 0 + 3B.$ (1)

Using the laws we exemplified in parts (a) through (d), (1) becomes $2A + C = 3B$. Now,

$$\frac{1}{3}(2A + C) = \frac{1}{3}(3B)$$

$$B = \frac{1}{3}(2A + C).$$

Now

$$2A + C = \begin{bmatrix} 2(-1) & 2(3) \\ 2(0) & 2(0) \end{bmatrix} + \begin{bmatrix} -2 & -1 \\ -1 & 1 \end{bmatrix} = \begin{bmatrix} -4 & 5 \\ -1 & 1 \end{bmatrix}.$$

Therefore,

$$B = \frac{1}{3}(2A + C) = \frac{1}{3}\begin{bmatrix} -4 & 5 \\ -1 & 1 \end{bmatrix} = \begin{bmatrix} -\dfrac{4}{3} & \dfrac{5}{3} \\ -\dfrac{1}{3} & \dfrac{1}{3} \end{bmatrix}.$$

If $A = \begin{bmatrix} 1 & 2 \\ -1 & 3 \end{bmatrix}$ and $B = \begin{bmatrix} 2 & 1 \\ 0 & 1 \end{bmatrix}$, show $AB \neq BA$.

A is 2×2 and B is 2×2; the product AB is a 2×2 matrix.

$$AB = \begin{bmatrix} 1 & 2 \\ -1 & 3 \end{bmatrix}\begin{bmatrix} 2 & 1 \\ 0 & 1 \end{bmatrix} = \begin{bmatrix} 1 \times 2 + 2 \times 0 & 1 \times 1 + 2 \times 1 \\ -1 \times 2 + 3 \times 0 & -1 \times 1 + 3 \times 1 \end{bmatrix}$$

$$= \begin{bmatrix} 2+0 & 1+2 \\ -2+0 & -1+3 \end{bmatrix} = \begin{bmatrix} 2 & 3 \\ -2 & 2 \end{bmatrix}.$$

Now,

$$BA = \begin{bmatrix} 2 & 1 \\ 0 & 1 \end{bmatrix}\begin{bmatrix} 1 & 2 \\ -1 & 3 \end{bmatrix} = \begin{bmatrix} 2 \times 1 + 1 \times (-1) & 2 \times 2 + 1 \times 3 \\ 0 \times 1 + 1 \times (-1) & 0 \times 2 + 1 \times 3 \end{bmatrix}$$

$$= \begin{bmatrix} 2-1 & 4+3 \\ 0-1 & 0+3 \end{bmatrix} = \begin{bmatrix} 1 & 7 \\ -1 & 3 \end{bmatrix}$$

Thus, $AB \neq BA$.

Find $A(B + C)$ and $AB + AC$ if

$$A = \begin{bmatrix} 2 & 2 & 3 \\ 3 & -1 & 2 \end{bmatrix}, \quad B = \begin{bmatrix} 1 & 0 \\ 2 & 2 \\ 3 & -1 \end{bmatrix}$$

and $C = \begin{bmatrix} -1 & 2 \\ 1 & 0 \\ 2 & -2 \end{bmatrix}$.

$$B+C = \begin{bmatrix} 1 & 0 \\ 2 & 2 \\ 3 & -1 \end{bmatrix} + \begin{bmatrix} -1 & 2 \\ 1 & 0 \\ 2 & -2 \end{bmatrix}$$

$$\begin{bmatrix} 1+(-1) & 0+2 \\ 2+1 & 2+0 \\ 3+2 & -1+(-2) \end{bmatrix} = \begin{bmatrix} 0 & 2 \\ 3 & 2 \\ 5 & -3 \end{bmatrix}$$

then,

$$A(B+C) = \begin{bmatrix} 2 & 2 & 3 \\ 3 & -1 & 2 \end{bmatrix} \begin{bmatrix} 0 & 2 \\ 3 & 2 \\ 5 & -3 \end{bmatrix}$$

$$= \begin{bmatrix} 2\times0+2\times3+3\times5 & 2\times2+2\times2+3\times(-3) \\ 3\times0+(-1)\times3+2\times5 & 3\times2+(-1)\times2+2\times(-3) \end{bmatrix}.$$

$$A(B+C) = \begin{bmatrix} 0+6+15 & 4+4-9 \\ 0-3+10 & 6-2-6 \end{bmatrix} = \begin{bmatrix} 21 & -1 \\ 7 & -2 \end{bmatrix}.$$

$$AB = \begin{bmatrix} 2 & 2 & 3 \\ 3 & -1 & 2 \end{bmatrix} \begin{bmatrix} 1 & 0 \\ 2 & 2 \\ 3 & -1 \end{bmatrix}$$

$$= \begin{bmatrix} 2\times1+2\times2+3\times3 & 2\times0+2\times2+3\times(-1) \\ 3\times1+(-1)\times2+2\times3 & 3\times0+(-1)\times2+2\times(-1) \end{bmatrix}$$

$$= \begin{bmatrix} 2+4+9 & 0+4+(-3) \\ 3-2+6 & 0-2-2 \end{bmatrix} = \begin{bmatrix} 15 & 1 \\ 7 & -4 \end{bmatrix}.$$

$$AC = \begin{bmatrix} 2 & 2 & 3 \\ 3 & -1 & 2 \end{bmatrix} \begin{bmatrix} -1 & 2 \\ 1 & 0 \\ 2 & -2 \end{bmatrix}$$

$$= \begin{bmatrix} 2\times(-1)+2\times1+3\times2 & 2\times2+2\times0+3\times(-2) \\ 3\times(-1)+(-1)\times1+2\times2 & 3\times2+(-1)\times0+2\times(-2) \end{bmatrix}$$

$$AC = \begin{bmatrix} -2+2+6 & 4+0-6 \\ -3-1+4 & 6+0-4 \end{bmatrix} = \begin{bmatrix} 6 & -2 \\ 0 & 2 \end{bmatrix}.$$

Then,

$$AB + AC = \begin{bmatrix} 15 & 1 \\ 7 & -4 \end{bmatrix} + \begin{bmatrix} 6 & -2 \\ 0 & 2 \end{bmatrix} = \begin{bmatrix} 15+6 & 1+(-2) \\ 7+0 & -4+2 \end{bmatrix}$$

$$= \begin{bmatrix} 21 & -1 \\ 7 & -2 \end{bmatrix}$$

Remark:

$$A(B+C) = \begin{bmatrix} 21 & -1 \\ 7 & -2 \end{bmatrix} \text{ and } AB+AC = \begin{bmatrix} 21 & -1 \\ 7 & -2 \end{bmatrix}.$$

Thus, $A(B + C) = AB + AC$. This is called the left distributive law.

9.2 Determinants

Associated with each n by n matrix A, there is a specific real number called the determinant of A and sometimes denoted by det (A). The usual way to calculate the determinant for a 2 by 2 matrix is described below.

If

$$A = \begin{bmatrix} a_{11} & a_{12} \\ a_{21} & a_{22} \end{bmatrix}, \text{ then det } (A) = a_{11}a_{22} - a_{12}a_{21}.$$

It is common to use vertical bars to represent determinants.

Then

$$\det (A) = \begin{vmatrix} a_{11} & a_{12} \\ a_{21} & a_{22} \end{vmatrix} = a_{11}a_{22} - a_{12}a_{21}.$$

Here are three important definitions concerning the calculation of determinants for a matrix A where

$$A = \begin{bmatrix} a_{11} & a_{12} & a_{13} & \cdots & a_{1n} \\ a_{21} & a_{22} & a_{23} & \cdots & a_{2n} \\ & & \vdots & & \\ a_{n1} & a_{n2} & a_{n3} & \cdots & a_{nn} \end{bmatrix}$$

(1) The minor, M_{ij}, of a_{ij} is the determinant of the $(n-1)$ by $(n-1)$ matrix obtained by deleting the ith row and j^{th} column of A.

(2) The cofactor, A_{ij}, of a_{ij} is

$$A_{ij} = (-1)^{i+j} M_{ij}$$

(3) The determinant of A is the sum of the n products formed by multiplying each entry in a single row (or column) by its cofactor.

Here is an example which illustrates this procedure.

EXAMPLE

$$\begin{vmatrix} 0 & 1 & 1 \\ 1 & 1 & 0 \\ 2 & -4 & 1 \end{vmatrix} = 1(-1)^3 \begin{vmatrix} 1 & 1 \\ -4 & 1 \end{vmatrix} + 1(-1)^4 \begin{vmatrix} 0 & 1 \\ 2 & 1 \end{vmatrix} + 0(-1)^5 \begin{vmatrix} 0 & 1 \\ 2 & -4 \end{vmatrix}$$

$$= (-1)5 + 1(-2) + 0(-2)$$

$$= -7$$

In this instance, the determinant was determined by expanding about the second row. Of course, the determinant could be determined by expanding about any row or column.

Determinants have some properties that are useful because they simplify the calculations. Here are these properties.

(1) If each element in any row or each element in any column is 0, then the determinant is equal to 0.

(2) If any two rows or any two columns of a determinant are interchanged, then the resulting determinant is the additive inverse of the original determinant.

(3) If two rows (or columns) in a determinant are equal, then the determinant is equal to 0.

(4) If each entry of one row (or column) is multiplied by a real number c and added to the corresponding entry in another row (or column) in the determinant, then the resulting determinant is equal to the original determinant.

(5) If each entry in a row (or column) of a determinant is multiplied by a real number c, then the determinant is multiplied by c.

Here is an example which illustrates using these last two properties.

EXAMPLE

$$\begin{vmatrix} 15 & 1 & 13 \\ 18 & 1 & 16 \\ 21 & -1 & 21 \end{vmatrix} = 3\begin{vmatrix} 5 & 1 & 13 \\ 6 & 1 & 16 \\ 7 & -1 & 21 \end{vmatrix}$$

$$= 3\begin{vmatrix} 5+7 & 1+(-1) & 13+21 \\ 6+7 & 1+(-1) & 16+21 \\ 7 & -1 & 21 \end{vmatrix}$$

$$= 3\begin{vmatrix} 12 & 0 & 34 \\ 13 & 0 & 37 \\ 7 & -1 & 21 \end{vmatrix}$$

$$= 3\left(0(-1)^3\begin{vmatrix} 13 & 37 \\ 7 & 21 \end{vmatrix} + 0(-1)^4\begin{vmatrix} 12 & 34 \\ 7 & 21 \end{vmatrix} + (-1)(-1)^5\begin{vmatrix} 12 & 34 \\ 13 & 37 \end{vmatrix} \right)$$

$$= 3[0 + 0 + (12 \cdot 37 - 13 \cdot 34)]$$

$$= 3 \times 2$$

$$= 6$$

Problem Solving Example:

Compute the determinants of each of the following matrices and find which of the matrices are invertible.

(a) $\begin{bmatrix} 3 & 1 & 2 \\ 1 & 0 & 6 \\ -1 & 1 & 1 \end{bmatrix}$

(b) $\begin{bmatrix} -1 & 1 & 3 \\ 2 & 1 & 1 \\ 4 & 2 & 2 \end{bmatrix}$

(c) $\begin{bmatrix} 2 & 1 & 1 \\ 0 & 0 & 0 \\ 4 & 3 & 1 \end{bmatrix}$

 We can evaluate determinants by using the basic properties of the determinant function.

Properties of Determinants:

(1) If each element in a row (or column) is zero, the value of the determinant is zero.

(2) If two rows (or columns) of a determinant are identical, the value of the determinant is zero.

(3) The determinant of a matrix A and its transpose A^t are equal:(A^t has its rows equal to the columns of A and its columns equal to the rows of A, in order.)

$$|A| = |A^t|.$$

(4)   The matrix A has an inverse if and only if det $(A) \neq 0$.

$$\det(A) = \begin{vmatrix} 3 & 1 & 2 \\ 1 & 0 & 6 \\ -1 & 1 & 1 \end{vmatrix}$$

$$= 3\begin{vmatrix} 0 & 6 \\ 1 & 1 \end{vmatrix} - 1\begin{vmatrix} 1 & 6 \\ -1 & 1 \end{vmatrix} + 2\begin{vmatrix} 1 & 0 \\ -1 & 1 \end{vmatrix}$$

$$= 3(0-6) - 1(1+6) + 2(1-0)$$

$$= -18 - 7 + 2 = -23.$$

Since det $(A) = -23 \neq 0$, this matrix is invertible.

(b)

$$A = \begin{bmatrix} -1 & 1 & 3 \\ 2 & 1 & 1 \\ 4 & 2 & 2 \end{bmatrix}$$

Here det $(A) = 0$, since the third row is a multiple of the second row. Since det $(A) = 0$, the matrix is not invertible.

(c)

$$A = \begin{bmatrix} 2 & 1 & 1 \\ 0 & 0 & 0 \\ 4 & 3 & 1 \end{bmatrix}$$

$$\det (A) = \begin{vmatrix} 2 & 1 & 1 \\ 0 & 0 & 0 \\ 4 & 3 & 1 \end{vmatrix} = 0.$$

Here, each element in the second row is zero, therefore, the value of the determinant is zero. Since det $A = 0$, the matrix is not invertible.

9.3   The Inverse of a Square Matrix

For a given n by n matrix A, if there is an n by n matrix A^{-1} with the property that $AA^{-1} = A^{-1}A = I$, then A^{-1} is the multiplicative inverse of A. This section is devoted to determining when a square matrix has an inverse and the procedures for finding inverses, if they exist. Consider the square matrix A where

$$A = \begin{bmatrix} a_{11} & a_{12} & \cdots & a_{1n} \\ a_{21} & a_{22} & \cdots & a_{2n} \\ & & \vdots & \\ a_{n1} & a_{n2} & \cdots & a_{nn} \end{bmatrix}$$

If det (A) is the determinant for A and det $(A) \neq 0$, then A has an inverse A^{-1} and

$$A^{-1} = \frac{1}{\det(A)} \begin{bmatrix} A_{11} & A_{12} & \cdots & A_{n1} \\ A_{21} & A_{22} & \cdots & A_{n2} \\ & & \vdots & \\ A_{1n} & A_{2n} & \cdots & A_{nn} \end{bmatrix},$$

where A_{ij} is the cofactor of a_{ij}. If $\delta(A) = 0$, then A has no inverse. While this generalization describes a definite process for determining inverses, the process is very laborious for large matrices and, thus, is not very practical.

A procedure for determining inverses which is much less demanding, from a computational standpoint, makes use of what is known as elementary row operations. Here are the elementary row operations for an m by n matrix A.

(1) Multiply each element in a row of A by a nonzero real number c.

(2) Interchange any two rows of A.

(3) Multiply the entries of any row of A by a real number c and add to the corresponding entries of any other row.

If B is a matrix resulting from a finite number of elementary row operations on A, then A and B are row-equivalent matrices, and this is denoted by $A \sim B$. If the inverse of an n by n matrix A exists, then it can be obtained by using elementary row operations. The following result illustrates how this can be done.

If A has an inverse and if $[A|I]$ is the n by $2n$ matrix obtained by adjoining the n by n identity matrix to A, then

$$[A|I] \sim [I|A^{-1}].$$

Here is an example which illustrates how this theorem can be applied.

Given

$$A = \begin{vmatrix} 5 & 0 & 2 \\ 2 & 2 & 1 \\ -3 & 1 & -1 \end{vmatrix}$$

find A^{-1} using elementary row operations.

$$\begin{bmatrix} 5 & 0 & 2 & | & 1 & 0 & 0 \\ 2 & 2 & 1 & | & 0 & 1 & 0 \\ -3 & 1 & -1 & | & 0 & 0 & 1 \end{bmatrix}$$

$$\sim \begin{bmatrix} 5+(-2)2 & 0+(-2)2 & 2+(-2)1 & | & 1+(-2)0 & 0+(-2)1 & 0+(-2)0 \\ 2 & 2 & 1 & | & 0 & 1 & 0 \\ -3 & 1 & -1 & | & 0 & 0 & 1 \end{bmatrix}$$

$$= \begin{bmatrix} 1 & -4 & 0 & | & 1 & -2 & 0 \\ 2 & 2 & 1 & | & 0 & 1 & 0 \\ -3 & 1 & -1 & | & 0 & 0 & 1 \end{bmatrix}$$

$$\sim \begin{bmatrix} 1 & -4 & 0 & | & 1 & -2 & 0 \\ 2+(-2)1 & 2+(-2)(-4) & 1+(-2)0 & | & 0+(-2)1 & 1+(-2)(-2) & 0+(-2)0 \\ -3+3(1) & 1+3(-4) & -1+3(0) & | & 0+3(1) & 0+3(-2) & 1+3(0) \end{bmatrix}$$

$$= \begin{bmatrix} 1 & -4 & 0 & | & 1 & -2 & 0 \\ 0 & 10 & 1 & | & -2 & 5 & 0 \\ 0 & -11 & -1 & | & 3 & -6 & 1 \end{bmatrix}$$

$$\sim \begin{bmatrix} 1 & -4 & 0 & | & 1 & -2 & 0 \\ 0+0 & 10+(-11) & 1+(-1) & | & -2+3 & 5+(-6) & 0+1 \\ 0 & -11 & -1 & | & 3 & -6 & 1 \end{bmatrix}$$

$$= \begin{bmatrix} 1 & -4 & 0 & 1 & -2 & 0 \\ 0 & -1 & 0 & 1 & -1 & 1 \\ 0 & -11 & -1 & 3 & -6 & 1 \end{bmatrix}$$

$$\sim \begin{bmatrix} 1 & -4 & 0 & 1 & -2 & 0 \\ 0 & -1 & 0 & 1 & -1 & 1 \\ 0 & -11 & -1 & 3 & -6 & 1 \end{bmatrix}$$

$$\sim \begin{bmatrix} 1+4(0) & -4+4(1) & 0+4(0) & 1+4(-1) & -2+4(1) & 0+4(-1) \\ 0 & 1 & 0 & -1 & 1 & -1 \\ 0+11(0) & -11+11(1) & -1+11(0) & 3+11(-1) & -6+11(1) & 1+11(-1) \end{bmatrix}$$

$$= \begin{bmatrix} 1 & 0 & 0 & -3 & 2 & -4 \\ 0 & 1 & 0 & -1 & 1 & -1 \\ 0 & 0 & -1 & -8 & 5 & -10 \end{bmatrix}$$

$$\sim \begin{bmatrix} 1 & 0 & 0 & -3 & 2 & -4 \\ 0 & 1 & 0 & -1 & 1 & -1 \\ 0 & 0 & 1 & 8 & -5 & 10 \end{bmatrix}$$

Thus,

$$A^{-1} = \begin{bmatrix} -3 & 2 & -4 \\ -1 & 1 & -1 \\ 8 & -5 & 10 \end{bmatrix}$$

Find the inverses of the following matrices.

(a) $A = \begin{bmatrix} 3 & 1 \\ -1 & 6 \end{bmatrix}$

(b) $A = \begin{bmatrix} 1 & -7 & -14 \\ 2 & 1 & -1 \\ 1 & 3 & 4 \end{bmatrix}$

(c) $A = \begin{bmatrix} 3 & 1 & 0 \\ 1 & -1 & 2 \\ 1 & 1 & 1 \end{bmatrix}$.

The method of solution is the same in all three cases, namely, forming the block matrix $[A|I]$ where I is the $n \times n$ identity matrix, and using elementary row operations to reduce it to $[I|A^{-1}]$.

(a)

$$A = \begin{bmatrix} 3 & 1 \\ -1 & 6 \end{bmatrix}.$$

Now

$$[A:I] = \begin{bmatrix} 3 & 1 & 1 & 0 \\ -1 & 6 & 0 & 1 \end{bmatrix}.$$

Multiply the first row by 6:

$$\begin{bmatrix} 18 & 6 & 6 & 0 \\ -1 & 6 & 0 & 1 \end{bmatrix}$$

Subtract the second row from the first row:

$$\begin{bmatrix} 19 & 0 & 6 & -1 \\ -1 & 6 & 0 & 1 \end{bmatrix}$$

Multiply the second row by 19:

$$\begin{bmatrix} 19 & 0 & 6 & 1 \\ -19 & 114 & 0 & 19 \end{bmatrix}$$

Add the first row to the second row:

$$\begin{bmatrix} 19 & 0 & 6 & 1 \\ 0 & 114 & 0 & 18 \end{bmatrix}$$

Divide the first and second rows by 19:

$$\begin{bmatrix} 1 & 0 & \dfrac{6}{19} & \dfrac{-1}{19} \\ 0 & 6 & \dfrac{6}{19} & \dfrac{18}{19} \end{bmatrix}$$

Divide the second row by 6:

$$\begin{bmatrix} 1 & 0 & \dfrac{6}{19} & \dfrac{-1}{19} \\ 0 & 1 & \dfrac{1}{19} & \dfrac{3}{19} \end{bmatrix}$$

Therefore

$$A^{-1} = \begin{bmatrix} \dfrac{6}{19} & \dfrac{-1}{19} \\ \dfrac{1}{19} & \dfrac{3}{19} \end{bmatrix}$$

(b)

$$A = \begin{bmatrix} 1 & -7 & -14 \\ 2 & 1 & -1 \\ 1 & 3 & 4 \end{bmatrix}$$

$$[A|\,I] = \begin{bmatrix} 1 & -7 & -14 & 1 & 0 & 0 \\ 2 & 1 & -1 & 0 & 1 & 0 \\ 1 & 3 & 4 & 0 & 0 & 1 \end{bmatrix}$$

Subtract the first row from the third row:

$$\begin{bmatrix} 1 & -7 & -14 & 1 & 0 & 0 \\ 2 & 1 & -1 & 0 & 1 & 0 \\ 0 & 10 & 18 & -1 & 0 & 1 \end{bmatrix}$$

Divide the third row by 2:

$$\begin{bmatrix} 1 & -7 & -14 & 1 & 0 & 0 \\ 2 & 1 & -1 & 0 & 1 & 0 \\ 0 & 5 & 9 & -\frac{1}{2} & 0 & \frac{1}{2} \end{bmatrix}$$

Add −2 times the first row to the second row:

$$\begin{bmatrix} 1 & -7 & 9 & 1 & 0 & 0 \\ 0 & 15 & 27 & -2 & 1 & 0 \\ 0 & 5 & 9 & -\frac{1}{2} & 0 & \frac{1}{2} \end{bmatrix}$$

Divide the second row by 3:

$$\begin{bmatrix} 1 & -7 & -14 & 1 & 0 & 0 \\ 0 & 5 & 9 & -\frac{2}{3} & \frac{1}{3} & 0 \\ 0 & 5 & 9 & -\frac{1}{2} & 0 & \frac{1}{2} \end{bmatrix}$$

Subtract the second row from the third row:

$$\begin{bmatrix} 1 & -7 & -14 & 1 & 0 & 0 \\ 0 & 5 & 9 & -\frac{2}{3} & \frac{1}{3} & 0 \\ 0 & 0 & 0 & \frac{1}{6} & -\frac{1}{3} & \frac{1}{3} \end{bmatrix}$$

At this point A is row equivalent to

$$F = \begin{bmatrix} 1 & -7 & -14 \\ 0 & 5 & 9 \\ 0 & 0 & 0 \end{bmatrix}$$

The matrix A is singular (det $A = 0$) and therefore A does not have an inverse.

(c)

$$A = \begin{bmatrix} 3 & 1 & 0 \\ 1 & -1 & 2 \\ 1 & 1 & 1 \end{bmatrix}$$

$$[A:I] = \begin{bmatrix} 3 & 1 & 0 & 1 & 0 & 0 \\ 1 & -1 & 2 & 0 & 1 & 0 \\ 1 & 1 & 1 & 0 & 0 & 1 \end{bmatrix}$$

Interchange the first and third rows:

$$\begin{bmatrix} 1 & 1 & 1 & 0 & 0 & 1 \\ 1 & -1 & 2 & 0 & 1 & 0 \\ 3 & 1 & 0 & 1 & 0 & 0 \end{bmatrix}$$

Subtract the first row from the second row and add −3 times the first row to the third row:

$$\begin{bmatrix} 1 & 1 & 1 & 0 & 0 & 1 \\ 0 & -2 & 1 & 0 & 1 & -1 \\ 0 & -2 & -3 & 1 & 0 & -3 \end{bmatrix}$$

Divide the second row by −2:

$$\begin{bmatrix} 1 & 1 & 1 & 0 & 0 & 1 \\ 0 & 1 & -\frac{1}{2} & 0 & -\frac{1}{2} & \frac{1}{2} \\ 0 & -2 & -3 & 1 & 0 & -3 \end{bmatrix}$$

Subtract the second row from the first row:

$$\begin{bmatrix} 1 & 0 & \frac{3}{2} & 0 & \frac{1}{2} & \frac{1}{2} \\ 0 & 1 & -\frac{1}{2} & 0 & -\frac{1}{2} & \frac{1}{2} \\ 0 & -2 & -3 & 1 & 0 & -3 \end{bmatrix}$$

Add 2 times the second row to the third row:

$$\begin{bmatrix} 1 & 0 & \frac{3}{2} & 0 & \frac{1}{2} & \frac{1}{2} \\ 0 & 1 & -\frac{1}{2} & 0 & -\frac{1}{2} & \frac{1}{2} \\ 0 & 0 & -4 & 1 & -1 & -2 \end{bmatrix}$$

Divide the third row by −4:

$$\begin{bmatrix} 1 & 0 & \frac{3}{2} & 0 & \frac{1}{2} & \frac{1}{2} \\ 0 & 1 & -\frac{1}{2} & 0 & -\frac{1}{2} & \frac{1}{2} \\ 0 & 0 & 1 & -\frac{1}{4} & \frac{1}{4} & \frac{2}{4} \end{bmatrix}$$

Add $-\dfrac{3}{2}$ times the third row to the first row and add $\dfrac{1}{2}$ times the third row to the second row:

$$\begin{bmatrix} 1 & 0 & 0 & \frac{3}{8} & \frac{1}{8} & -\frac{2}{8} \\ 0 & 1 & 0 & -\frac{1}{8} & -\frac{3}{8} & \frac{6}{8} \\ 0 & 0 & 1 & -\frac{1}{4} & \frac{1}{4} & \frac{2}{4} \end{bmatrix}$$

Thus

$$A^{-1} = \begin{bmatrix} \frac{3}{8} & \frac{1}{8} & -\frac{2}{8} \\ -\frac{1}{8} & -\frac{3}{8} & \frac{6}{8} \\ -\frac{1}{4} & \frac{1}{4} & \frac{2}{4} \end{bmatrix} = \frac{1}{8}\begin{bmatrix} 3 & 1 & -2 \\ -1 & -3 & 6 \\ -2 & 2 & 4 \end{bmatrix}$$

Problem Solving Example:

 Let

$$A = \begin{bmatrix} 1 & 2 \\ 3 & 4 \end{bmatrix}$$

Find the inverse of A directly by solving for the entries of the matrix B which satisfies the equation

$$A \times B = I,$$

where $A \times B$ is matrix multiplication.

 The problem asks us to solve for the entries $a, b, c,$ and d of the matrix

$$B = \begin{bmatrix} a & b \\ c & d \end{bmatrix},$$

given that $A \times B = I$.

Since $A \times B = I$, we have

$$\begin{bmatrix} 1 & 2 \\ 3 & 4 \end{bmatrix} \begin{bmatrix} a & b \\ c & d \end{bmatrix} = \begin{bmatrix} 1 & 0 \\ 0 & 1 \end{bmatrix}$$

After performing the multiplication of the matrices, we obtain

$$\begin{bmatrix} a+2c & b+2d \\ 3a+4c & 3b+4d \end{bmatrix} = \begin{bmatrix} 1 & 0 \\ 0 & 1 \end{bmatrix}.$$

Recall that two matrices are equal if, and only, if, their corresponding entries are equal. Thus from the last equation we may conclude that

$$a+2c = 1, \quad b+2d = 0, \quad 3a+4c = 0, \quad 3b+4d = 1.$$

From these four equations, we can obtain two sets of linear equations from which we can solve for each of a, b, c, d. That is, we have the set:

$$a + 2c = 1$$
$$3a + 4c = 0$$

whose solutions are $a = -2$ and $c = \dfrac{3}{2}$, and the set

$$b + 2d = 0$$
$$3b + 4d = 1$$

whose solutions are $b = 1$ and $d = \dfrac{-1}{2}$ Hence,

$$B = \begin{bmatrix} a & b \\ c & d \end{bmatrix} = \begin{bmatrix} -2 & 1 \\ \frac{3}{2} & \frac{-1}{2} \end{bmatrix}.$$

Since B satisfies $AB = I$, $B = A^{-1}$. Hence,

$$A^{-1} = \begin{bmatrix} -2 & 1 \\ \frac{3}{2} & -\frac{1}{2} \end{bmatrix}.$$

The method used consisted of obtaining sets of linear equations whose unique solutions yielded the required inverse.

Quiz: Matrices and Determinants

1. Find x if the determinant of the matrix $\mathbf{A}$ is $2x - 4$, where

$$A = \begin{bmatrix} x-2 & 0 \\ 0 & x-3 \end{bmatrix}$$

   (A) $x = 2$ or $x = 5$.   (D) $x = 2$ or $x = 3$.

   (B) $x = 1$ or $x = -1$.   (E) $x = -2$ or $x = 1$.

   (C) $x = 0$ or $x = 2$.

2. What is $3\begin{bmatrix} 6 & 9 \\ 10 & 12 \end{bmatrix}$?

   (A) $\begin{bmatrix} 9 & 12 \\ 13 & 5 \end{bmatrix}$.

   (B) $\begin{bmatrix} 6 & 27 \\ 30 & 18 \end{bmatrix}$.

   (C) $\begin{bmatrix} 18 & 0 \\ 0 & 6 \end{bmatrix}$.

   (D) $\begin{bmatrix} 9 & 12 \\ 13 & 15 \end{bmatrix}$.

   (E) None of the above.

3. Let $A = \begin{bmatrix} 3 & -1 & 0 \\ 2 & 1 & 9 \end{bmatrix}$ and $B = \begin{bmatrix} 1 & 0 & -5 \\ 3 & 1 & 2 \end{bmatrix}$. What is $4A - B$?

(A) $\begin{bmatrix} 11 & -4 & 5 \\ -1 & 0 & 7 \end{bmatrix}$.

(B) $\begin{bmatrix} 11 & -4 & 5 \\ 5 & 3 & 34 \end{bmatrix}$.

(C) $\begin{bmatrix} 8 & -4 & 20 \\ -4 & 0 & 28 \end{bmatrix}$.

(D) $\begin{bmatrix} 13 & -4 & -5 \\ 11 & 5 & 38 \end{bmatrix}$.

(E) $\begin{bmatrix} 2 & 0 & -5 \\ -1 & 0 & 7 \end{bmatrix}$.

4. Calculate $\begin{bmatrix} 1 & 0 \\ 1 & 2 \end{bmatrix} + \begin{bmatrix} 2 & 4 \\ 3 & 5 \end{bmatrix}$

(A) $\begin{bmatrix} -1 & -4 \\ -2 & -3 \end{bmatrix}$.

(B) $\begin{bmatrix} 3 & 1 \\ 1 & 7 \end{bmatrix}$.

(C) $\begin{bmatrix} 3 & 4 \\ 4 & 7 \end{bmatrix}$.

(D) $\begin{bmatrix} 2 & 0 \\ 3 & 10 \end{bmatrix}$.

(E) $\begin{bmatrix} 1 & 6 \\ 3 & 8 \end{bmatrix}$.

5. What is the determinant of $\begin{bmatrix} 15 & 2 \\ 21 & 4 \end{bmatrix}$?

(A) 102.   (D) 81.

(B) 3.   (E) 6.

(C) 18.

6. What is $\begin{vmatrix} 0 & 1 & 2 \\ 2 & 3 & 4 \\ 5 & 0 & 1 \end{vmatrix}$?

(A) −12.   (D) 32.

(B) 0.   (E) −18.

(C) −20.

7. What is the inverse matrix of $\begin{bmatrix} 5 & 0 \\ 2 & 4 \end{bmatrix}$?

(A) $\begin{bmatrix} 1 & 0 \\ 0 & 1 \end{bmatrix}$.

(B) $\begin{bmatrix} \dfrac{1}{5} & 0 \\ -\dfrac{1}{10} & \dfrac{1}{4} \end{bmatrix}$.

(C) $\begin{bmatrix} \dfrac{1}{2} & \dfrac{1}{4} \\ \dfrac{1}{5} & \dfrac{1}{10} \end{bmatrix}$.

(D) $\begin{bmatrix} 1 & 0 \\ 0 & \dfrac{1}{2} \end{bmatrix}$.

(E) $\begin{bmatrix} 10 & 0 \\ 1 & 1 \end{bmatrix}$.

8. What is $\begin{bmatrix} 2 & -1 \\ 0 & 3 \end{bmatrix} + \begin{bmatrix} 6 & -5 \\ 1 & -2 \end{bmatrix}\begin{bmatrix} 4 & 1 \\ 7 & -3 \end{bmatrix}$?

(A) $\begin{bmatrix} -12 & 35 \\ -30 & 21 \end{bmatrix}$.

(B) $\begin{bmatrix} 12 & -5 \\ 8 & -2 \end{bmatrix}$.

(C) $\begin{bmatrix} -9 & 20 \\ -10 & 10 \end{bmatrix}$.

(D) $\begin{bmatrix} -10 & 26 \\ 11 & -2 \end{bmatrix}$.

(E) $\begin{bmatrix} 2 & -4 \\ -6 & -1 \end{bmatrix}$.

9. Let $A = \begin{bmatrix} 1 & 0 & 1 \\ 6 & 2 & 9 \\ -3 & 2 & 0 \end{bmatrix}$, $B = \begin{bmatrix} -2 \\ 1 \\ 3 \end{bmatrix}$, and $C = \begin{bmatrix} x \\ y \\ z \end{bmatrix}$

Find C if $A^{-1}B = C$.

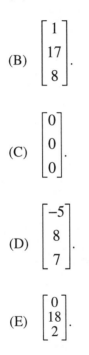

(A) No solution.

(B) $\begin{bmatrix} 1 \\ 17 \\ 8 \end{bmatrix}$.

(C) $\begin{bmatrix} 0 \\ 0 \\ 0 \end{bmatrix}$.

(D) $\begin{bmatrix} -5 \\ 8 \\ 7 \end{bmatrix}$.

(E) $\begin{bmatrix} 0 \\ 18 \\ 2 \end{bmatrix}$.

10. Find a matrix B such that $A^{-1}B = I$. when

$$A^{-1} = \frac{1}{47}\begin{bmatrix} 6 & 19 & 4 \\ 3 & -14 & 2 \\ -5 & -8 & 9 \end{bmatrix}$$

(A) $47\begin{bmatrix} 6 & 19 & 4 \\ 3 & -14 & 2 \\ -5 & -8 & 9 \end{bmatrix}.$

(B) $\begin{bmatrix} 110 & 203 & -2 \\ 1 & -2 & 0 \\ 2 & 1 & 3 \end{bmatrix}.$

(C) $\begin{bmatrix} 11 & 20 & -94 \\ 1 & -2 & 7 \\ 94 & 47 & 141 \end{bmatrix}.$

(D) $\begin{bmatrix} 2 & 5 & -4 \\ 1 & -2 & 0 \\ 4 & 2 & 6 \end{bmatrix}.$

(E) None of the above.

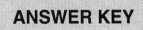

ANSWER KEY

1.	(A)	6.	(A)
2.	(E)	7.	(B)
3.	(B)	8.	(C)
4.	(C)	9.	(A)
5.	(C)	10.	(E)

CHAPTER 10

Miscellaneous Topics

10.1 The Fundamental Counting Principle, Permutations, and Combinations

If a first task can be performed in m ways, and if after the first task has been performed a second task can be performed in n ways, then the two tasks can be performed in mn ways. This is called the fundamental counting principle.

A permutation of r objects from a set of n objects is an ordered arrangement of those r objects. The symbol $_nP_r$ is used to represent the number of permutations of n objects taken r at a time, and when $n \geq r$.

$$_nP_r = \frac{n!}{(n-r)!}$$

A combination of r objects from a set of n objects is an unordered selection of r objects from the n objects. The symbols $\binom{n}{r}$ and $_nC_r$ are both used to represent the number of combinations of n objects taken r at a time and

$$_nC_r = \binom{n}{r} = \frac{n!}{r!(n-r)!}$$

Problem Solving Examples:

Q Find $_9P_4$.

A Using the general formula for permutations of n different things taken r at a time $_nP_r = \dfrac{n!}{(n-r)!}$, we substitute 9 for n and 4 for r. Hence $_9P_4 = \dfrac{9!}{(9-4)!} = \dfrac{9!}{5!}$. Evaluating our factorials, we obtain:

$$_9P_4 = \frac{9 \times 8 \times 7 \times 6 \times (5 \times 4 \times 3 \times 2 \times 1)}{(5 \times 4 \times 3 \times 2 \times 1)}$$

cancelling 5! in the numerator and denominator:

$$_9P_4 = 9 \times 8 \times 7 \times 6$$
$$= 3,024.$$

Q In how many ways may three books be placed next to each other on a shelf?

A We construct a pattern of three boxes to represent the places where the three books are to be placed next to each other on the shelf:

Since there are three books, the first place may be filled in three ways. There are then two books left, so that the second place may be filled in two ways. There is only one book left to fill the last place. Hence our boxes take the following form:

3	2	1

The Fundamental Principle of Counting states that if one thing can be done in a different ways and, when it is done in any one of these ways, a second thing can be done in b different ways, and a third thing can be done in c ways, . . . then all the things in succession can be done in $a \times b \times c$... different ways. Thus the books can be arranged in $3 \times 2 \times 1 = 6$ ways.

This can also be seen using the following approach. Since the arrangement of books on the shelf is ordered, this is a permutation problem. Recalling the general formula for the number of permutations of n things taken r at a time, $_nP_r = n!/(n-r)!$, we replace n by 3 and r by 3 to obtain

$$_3P_3 = \frac{3!}{(3-3)!} = \frac{3!}{0!} = \frac{3 \times 2 \times 1}{1} = 6$$

 How many different five-card hands can be obtained from a fifty-two card deck?

 Since the order of the cards is unimportant, we are dealing with a combination problem, as opposed to a permutation problem. Recall our general formula for combinations $_nC_r = \dfrac{n!}{r!(n-r)!}$.

We want to know how many 5 card hands can be obtained from a 52 card deck, hence $r = 5$, $n = 52$. Substituting 5 for r and 52 for n in our general formula, we obtain:

$$_{52}C_5 = \frac{52!}{5!(52-5)!} - \frac{52!}{5!(47)!}.$$

Recall that

$$n! = n \times (n-1)! = n \times (n-1) \times (n-2)! = n \times (n-1)(n-2)(n-3)!...$$

hence $52! = \dfrac{52 \times 51 \times 50 \times 49 \times 48 \times (47)!}{5!(47)!}$

Cancelling $(47)!$ from numerator and denominator, and evaluating 5!:

$$\frac{52 \times 51 \times 50 \times 49 \times 48}{5 \times 4 \times 3 \times 2 \times 1}.$$

Performing the necessary multiplications and divisions:

$$= 2,598,960.$$

Thus 2,598,960 different five-card hands can be obtained from a fifty-two card deck.

 How many telephone numbers of four different digits each can be made from the digits 0,1,2,3,4,5,6,7,8,9?

 A different arrangement of the same four digits produces a different telephone number. Since we are concerned with the order in which the digits appear, we are dealing with permutation.

There are ten digits to choose from and four different ones are to be chosen at a time. The general formula for the number of permutations of n things taken r at a time is

$$P(n,r) = \frac{n!}{(n-r)!}.$$

Here $n = 10$, $r = 4$, and the desired number is

$$P(10,4) = \frac{10!}{(10-4)!} = \frac{10!}{6!} = \frac{10 \times 9 \times 8 \times 7 \times 6!}{6!} = 5040$$

Thus 5040 telephone numbers of four different digits each can be made from the ten digits.

10.2 The Binomial Theorem

The Binomial Theorem involves the expansion of products of the form $(a + b)^n$, where n is a positive integer. When n and k are positive integers with $k \le n$, the symbol $\binom{n}{k}$ represents the number of combinations of n different things taken k at a time, and as indicated in the previous section,

$$\binom{n}{k} = \frac{n!}{k!(n-k)!}$$

The Binomial Theorem is as follows:

$$(a+b)^n = \binom{n}{0}a^n + \binom{n}{1}a^{n-1}b + \binom{n}{2}a^{n-2}b^2 + \ldots + \binom{n}{n-1}ab^{n-1} + \binom{n}{n}b^n$$

The example below illustrates the use of this famous theorem.

EXAMPLE

$$(2x-3y)^4 = \binom{4}{0}(2x)^4 + \binom{4}{1}(2x)^3(-3y) + \binom{4}{2}(2x)^2(-3y)^2$$

$$+ \binom{4}{3}(2x)(-3y)^3 + \binom{4}{4}(-3y)^4$$

$$= 16x^4 - 96x^3y + 216x^2y^2 - 216xy^3 + 81y^4$$

Problem Solving Example:

 What is the 9th term in the explanation of $(3x-y)^{14}$?

 Using the Binomial Theorem we may obtain the 9th term without expanding the entire expression. The 9th term will be:

$$\binom{14}{8}(3x)^6(-y)^8.$$

$\binom{14}{8} = 3003$ thus we have $(3003)(729)x^6y^8$.

Finally we find the 9th term to be $2189187x^6y^8$.

10.3 Sequences and Series

A sequence is a function whose domain is the set of all natural numbers. All the sequences described in this section will have a subset in the set of all real numbers as their range. It is common to let a_n represent the nth term of the sequence. For example, if

$$a_n = 10n$$

then the sequence is $10, 20, 30, 40 \ldots$.

The sum of the first n terms of the sequence,

$$a_1, a_2, a_3, \ldots a_n,$$

is indicated by

$$a_1 + a_2 + a_3 + \ldots + a_n,$$

and the sum of these terms is called a series. The Greek letter Σ is used to represent this sum as indicated below.

$$\sum_{k=1}^{n} a_k = a_1 + a_2 + a_3 + \ldots + a_n$$

For a fixed number a and a fixed number d, the sequence

$$a, a + d, a + 2d, a + 3d, \ldots$$

is called an arithmetic sequence, and the nth term of this sequence is given by

$$a_n = a + (n - 1)d$$

The number a is called the first term and d is called the common difference. The symbol S_n is used to represent the corresponding series and

$$\sum_{K=1}^{n} a+(K-1)d$$

or $S_n = n\left(\dfrac{a_1 + a_n}{2}\right)$

A sequence of the form

$a, ar, ar^2, \ldots$

is called a geometric sequence, where a is the first term and r is the common ratio. The nth term of such a sequence is

$a_n = ar^{n-1}$

The symbol S_n is used to represent the corresponding series and

$$S_n = \frac{a\left(1-r^n\right)}{1-r}.$$

Expressions of the form

$a + ar + ar^2 \ldots$

are called infinite geometric series. When $|r| < 1$, the sum S of the infinite geometric series exists and

$$S = \frac{a}{1-r}$$

Problem Solving Example:

In an arithmetic sequence, $a_1 = 29$ and $a_8 = 78$. Find the common difference d and the sixth term a_6.

The nth term in an arithmetic sequence is given by

$a_n = a + (n-1)d$ where a is the initial term and d is the common difference. Using the given information we can first find d.

$$78 = 29 + (8-1)d$$
$$78 = 29 + 7d$$
$$49 = 7d$$
$$7 = 9$$

Thus the common difference is 7. We may now use this information to obtain a_6.

$$a_6 = 29 + (6-1)(7)$$
$$= 29 + 35$$
$$= 64$$

thus the sixth term is 64.

10.4 Mathematical Induction

The material in this section refers to the set of all natural numbers N. Suppose A is a set with the properties

(1) $A \subseteq N$

(2) $1 \in A$ and

(3) if $k \in A$ then $k + 1 \in A$.

It can be concluded that in this case, $A = N$. This generalization provides the basis for proof by mathematical induction. The corresponding theorem is as follows:

If for a given statement about positive integers, the statement is true for 1 and if it is true for $n = k$, then it is true for $n = k + 1$, then the statement is true for all positive integers.

Here is an example which illustrates the use of this theorem.

EXAMPLE

Prove that

$$1+2+3+...+n = \frac{n(n+1)}{2}$$

Proof:

$$1 = \frac{1(1+1)}{2}$$

Now assume that the statement is true when $n = k$. Then

$$1+2+3+...+k = \frac{k(k+1)}{2}.$$

Then

$$1+2+3+...+k+(k+1) = \frac{k(k+1)}{2}+k+1$$
$$= \frac{k(k+1)+2(k+1)}{2}$$
$$= \frac{(k+1)(k+2)}{2}$$

Thus, the statement is true when $n = k + 1$. Thus, the statement is true for all positive integers.

Problem Solving Examples:

 Prove by mathematical induction

$$1^2 + 2^2 + 3^2 + ... + n^2 = \frac{1}{6}n(n+1)(2n+1)$$

 Mathematical induction is a method of proof. The steps are:

(1) The verification of the proposed formula or theorem for the smallest value of n. It is desirable, but not necessary, to verify it for several values of n.

(2) The proof that if the proposed formula or theorem is true for $n = k$, some positive integer, it is true also for $n = k + 1$. That is, if the proposition is true for any particular value of n, it must be true for the next larger value of n.

(3) A conclusion that the proposed formula holds true for all values of n.

Proof: Step 1. Verify:

For $n = 1$: $1^2 = \dfrac{1}{6}(1)(1+1)[2(1)+1] = \dfrac{1}{6}(1)(2)(3) = \dfrac{1}{6}(6) = 1$

$1 = 1$ ✔

For

$n = 2$: $1^2 + 2^2 = \dfrac{1}{6}(2)(2+1)[2(2)+1] = \dfrac{1}{6}(2)(3)(5) = \dfrac{1}{6}(6)(5)$

$1 + 4 = (1)(5)$ ✔

$\qquad 5 = 5$

For $n = 3$: $1^2 + 2^2 + 3^2 = \dfrac{1}{6}(3)(3+1)[2(3)+1]$

$1 + 4 + 9 = \dfrac{1}{6}(3)(4)(7) = \dfrac{1}{6}(12)(7) = 14$ ✔

$14 = 14$

Step 2. Let k represent any particular value of n. For $n = k$, the formula becomes

$$1^2 + 2^2 + 3^2 + \ldots + k^2 = \dfrac{1}{6}k(k+1)(2k+1) \qquad (1)$$

For $n = k + 1$, the formula is

$$1^2 + 2^2 + 3^2 + \ldots + k^2 + (k+1)^2 = \frac{1}{6}(k+1)\big[(k+1)+1\big]\big[2(k+1)+1\big]$$

$$= \frac{1}{6}(k+1)(k+2)(2k+3). \qquad (2)$$

We must show that if the formula is true for $n = k$, then it must be true for $n = k + 1$. In other words, we must show that (2) follows from (1). The left side of (1) can be converted into the left side of (2) by merely adding $(k+1)^2$. All that remains to be demonstrated is that when $(k+1)^2$ is added to the right side of (1), the result is the right side of (2).

$$1^2 + 2^2 + 3^2 + \ldots + k^2 + (k+1)^2 = \frac{1}{6}(k+1)(2k+1) + (k+1)^2$$

Factor out $(k + 1)$:

$$1^2 + 2^2 + 3^2 \ldots + k^2 + (k+1)^2 = (k+1)\left[\frac{1}{6}k(2k+1) + (k+1)\right]$$

$$= (k+1)\left[\frac{k(2k+1)}{6} + \frac{(k+1)6}{6}\right]$$

$$= (k+1)\frac{2k^2 + k + 6k + 6}{6}$$

$$= \frac{(k+1)\big(2k^2 + 7k + 6\big)}{6}$$

$$= \frac{1}{6}(k+1)(k+2)(2k+3),$$

since

$$2k^2 + 7k + 6 = (k+2)(2k+3).$$

Thus, we have shown that if we add $(k+1)^2$ to both sides of the equation for $n = k$, then we obtain the equation or formula for $n = k + 1$. We have thus established that if (1) is true, then (2) must be true; that is, if the formula is true for $n = k$, then it must be true for $n = k + 1$. In other words, we have proved that if the proposition is true for a certain positive integer k, then it is also true for the next greater integer $k + 1$.

Step 3. The proposition is true for $n = 1,2,3$ (Step 1). Since it is true for $n = 3$, it is true for $n = 4$ (Step 2, where $k = 3$ and $k + 1 = 4$). Since it is true for $n = 4$, it is true for $n = 5$, and so on, for all positive integers n.

 Prove:

$$1 \times 2 + 2 \times 3 + 3 \times 4 + \ldots + n(n+1) = \frac{n(n+1)(n+2)}{3}.$$

Solution by mathematical induction: The steps for a proof by mathematical induction are:

I)   check validity of formula for $n = 1$

II)   assume the formulation is true for $n = k$

III)   prove it is true for $n = k + 1$.

I. For $n = 1$ the formula gives

$$1(1+1) = 1(2) = 2 = \frac{1(1+1)(1+2)}{3} = \frac{1 \times 2 \times 3}{3} = 2$$

which is correct and completes Step I.

II) Assume the formula is true for $n = k$:

$$1 \times 2 + 2 \times 3 + 3 \times 4 + \ldots + k(k+1) = \frac{k(k+1)(k+2)}{3}.$$

Prove the formula is true for $n = k + 1$, that is, prove

$$1 \times 2 + 2 \times 3 + \ldots + k(k+1) + (k+1)(k+2) = \frac{(k+1)(k+2)(k+3)}{3},$$

$(k+1)(k+2)$ is added to both members of the first equation in this step; this gives

$$\left[1 \times 2 + 2 \times 3 + \ldots + k(k+1)\right] + (k+1)(k+2)$$

$$= \left[\frac{k(k+1)(k+2)}{3}\right] + (k+1)(k+2)$$

Factoring out $(k+1)(k+2)$

$$= (k+1)(k+2)\left(\frac{k}{3}+1\right)$$

$$= (k+1)(k+2)\left(\frac{k}{3}+1\right)\left(\frac{3}{3}\right)$$

$$= \frac{(k+1)(k+2)(k+3)}{3},$$

$$= \frac{(k+1)\left[(k+1)+1\right]\left[(k+2)+1\right]}{3}$$

which is of the same form as the result we assumed to be true for k terms, $k + 1$ taking the place of k. Since the statement is true for $n = 1$ and $n = k + 1$ assuming it was true for $n = k$, then the statement is true for all n.

 Prove by mathematical induction that
$$1 + 7 + 13 + \ldots + (6n-5) = n(3n-2).$$

 (1) The proposed formula is true for $n = 1$, since $1 = 1(3-2)$.

(2) Assume the formula to be true for $n = k$, a positive integer; that is, assume

$$1 + 7 + 13 + \ldots + (6k - 5) = k(3k - 2). \qquad (1)$$

Under this assumption we wish to show that

$$1 + 7 + 13 + \ldots + (6k - 5) + (6k + 1) = (k + 1)(3k + 1). \qquad (2)$$

When $(6k + 1)$ is added to both members of (1), we have on the right $k(3k - 2) + (6k + 1) = 3k^2 + 4k + 1 = (k + 1)(3k + 1)$;

hence, if the formula is true for $n = k$ it is true for $n = k + 1$.

(3) Since the formula is true for $n = k = 1$ (Step 1), it is true for $n = k + 1 = 2$; being true for $n = k = 2$ it is true for $n = k + 1 = 3$; and so one, for every positive integral value of n.

10.5 Zeros of a Polynomial

As indicated in Section 3.5, an expression of the form

$$a_n x^n + a_{n-1} x^{n-1} + \ldots + a_1 x + a_0$$

is a polynomial and a function of the form

$$P(x) = a_n x^n + a_{n-1} x^{n-1} + \ldots + a_1 x + a_0$$

is a polynomial function. A number r is said to be a zero of this polynomial provided that $P(r) = 0$. The Factor Theorem, which follows, illustrates the relationship between zeros and factors of a polynomial.

For a polynomial $P(x)$, r is a zero of $P(x)$ if, and only if, $x-r$ is a factor of $P(x)$. Here are two other theorems, which are useful in terms of finding zeros of polynomials.

If $P(x)$ is a polynomial, and if $P(a) < 0$ and $P(b) > 0$, then $P(x)$ has a zero between a and b.

If $\dfrac{p}{q}$ is a rational number in lowest terms which is zero of $P(x)$ where

$$P(x) = a_n x^n + a_{n-1} x^{n-1} + \ldots a_1 x + a_0$$

and where $a_0, a_1, a_2, \ldots$ and a_n are all integers, then p is a factor of a_0 and q is a factor of a_n.

Here is an example which illustrates the use of these theorems.

EXAMPLE

Find the zeros of

$$f(x) = 2x^4 - 3x^3 - 7x^2 - 8x + 6.$$

Then the possible rational zeros are

$\pm 1, \pm 2, \pm 3, \pm 6, \pm \dfrac{1}{2},$ and $\pm \dfrac{3}{2}.$ Applying standard division procedures as follows,

$$
\begin{array}{r}
2x^3 + 3x^2 + 2x - 2 \\
x - 3 \overline{\smash{\big)}\ 2x^4 - 3x^3 - 7x^2 - 8x + 6} \\
\underline{2x^4 - 6x^3} \\
3x^3 - 7x^2 \\
\underline{3x^3 - 9x^2} \\
2x^2 - 8x \\
\underline{2x^2 - 6x} \\
-2x + 6 \\
\underline{-2x + 6} \\
\end{array}
$$

Indicate that 3 is a zero of this polynomial and

$$2x^4 - 3x^3 - 7x^2 - 8x + 6 = (x - 3)(2x^3 + 3x^2 + 2x - 2).$$

Then

$$x - \frac{1}{2} \overline{\smash{\big)}\,2x^3 + 3x^2 + 2x - 2}$$

with quotient $2x^2 + 4x + 4$ and the long division steps:

$$\underline{2x^3 - x^2}$$
$$4x^2 + 2x$$
$$\underline{4x^2 - 2x}$$
$$4x - 2$$
$$\underline{4x - 2}$$

and $2x^4 - 3x^3 - 7x^2 - 8x + 6 = (x - 3)\left(x - \frac{1}{2}\right)\left(2x^2 + 4x + 4\right).$

Applying the quadratic formula to

$$2x^2 + 4x + 4 = 0$$

gives $-1 \pm i$ as solutions. Thus, the zeros of

$$2x^4 - 3x^3 - 7x^2 - 8x + 6$$

are $3, \frac{1}{2}, -1 + i,$ and $-1 - i$.

Next, consider the polynomial

$$f(x) = x^3 - x^2 - 3.$$

The only possible rational zeros of this polynomial are ± 1 and ± 3. However, none of these numbers are actually zeros of the polynomial, which means that this polynomial has no rational zeros. However, it is possible to approximate zeros of this polynomial. Since

$$f(1) = -3 \quad \text{and} \quad f(2) = 1,$$

$f(x)$ has a zero between 1 and 2. Also $f(1.8) = -0.408$ and $f(1.9) = 0.249$ so the polynomial has a zero between 1.8 and 1.9. This process can be continued so the zero could be approximated to any desired level of accuracy.

The procedures described in this section apply only to finding rational number zeros of polynomials. However, when a polynomial has no rational zeros, it is possible to approximate irrational zeros using the process described in the example above. A calculator or a computer is a vital tool in this process.

It is often very difficult to find non-real number zeros of polynomials. Here is one theorem concerning nonreal zeros.

If $\quad f(x) = a_n x^n + a_{n-1} x^{n-1} + \dots a_1 x + a_0 \quad$ and $\quad a_0, a_1, a_2, \dots, a_n$

are real numbers, and if $a + bi$ is a zero of $f(x)$, then $a - bi$ is a zero of $f(x)$.

This theorem is applied in the following example.

EXAMPLE

Find all the zeros of

$$f(x) = x^4 - 2x^3 + 3x^2 - 2x + 2$$

given that $1 + i$ is one zero. Since $1 + i$ is a zero, $1 - i$ is a zero, and there is a polynomial $Q(x)$ such that

$$P(x) = \left[x - (1+i)\right]\left[x - (1-i)\right]Q(x)$$
$$= \left(x^2 - 2x + 2\right)Q(x).$$

But since

$$\frac{x^4 - 2x^3 + 3x^2 - 2x + 2}{x^2 - 2x + 2} = x^2 + 1$$

$$P(x) = \left(x^2 - 2x + 2\right)\left(x^2 + 1\right)$$

Thus, the zeros of this polynomial are $1 + i$, $1 - i$, i, and $-i$.

Now consider the polynomial

$$R(x) = b_n x^n + b_{n-1} x^{n-1} + \dots + b_1 x + b_0$$

where b_0, b_1 . . . and b_n are complex numbers. Such a polynomial is said to be a polynomial over the complex numbers. Here is the fundamental theorem of algebra.

Every polynomial of degree $n \geq 1$ over the complex number has at least one complex number zero.

The factor theorem stated earlier in this section can be extended to polynomials over the complex numbers. Here is a related theorem.

Every polynomial of degree $n \geq 1$ over the complex numbers can be expressed as a product of a constant and linear factors of the form $(x-x_j)$ where x_j is a complex number.

If a factor $x-x_j$ occurs k times in a linear factorization of $P(x)$, then x_j is said to be a zero of multiplicity k. This means that every polynomial with complex number coefficients has exactly n zeros.

Problem Solving Examples:

 Locate the roots of $x^3 - 3x^2 - 6x + 9 = 0$

 If we let $f(x)$ be a function, then a solution of the equation $f(x) = 0$ is called a root of the equation.

In this particular case let the function $f(x) = x^3 - 3x^2 - 6x + 9$ and set it equal to zero to find its roots. When $f(x) = 0$, the graph of this equation crosses the x-axis. These x-values are the roots of the function.

To locate the roots of $x^3 - 3x^2 - 6x + 9 = 0$, we consider the function $y = x^3 - 3x^2 - 6x + 9$, assign consecutive integers from -3 to 5 to x, compute each corresponding value of y, and record the results.

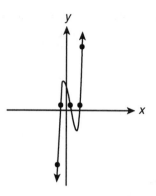

Table of Results

x	$x^3 - 3x^2 - 6x + 9$	=	y
-3	$(-3)^3 - 3(-3)^2 - 6(-3) + 9$	=	-27
-2	$(-2)^3 - 3(-2)^2 - 6(-2) + 9$	=	1
-1	$(-1)^3 - 3(-1)^2 - 6(-1) + 9$	=	11
0	$(0)^3 - 3(0)^2 - 6(0) + 9$	=	9
1	$(1)^3 - 3(1)^2 - 6(1) + 9$	=	1
2	$(2)^3 - 3(2)^2 - 6(2) + 9$	=	-7
3	$(3)^3 - 3(3)^2 - 6(3) + 9$	=	-9
4	$(4)^3 - 3(4)^2 - 6(4) + 9$	=	1
5	$(5)^3 - 3(5)^2 - 6(5) + 9$	=	29

Since $f(-3) = -27$ and $f(-2) = 1$, there is an odd number of roots between $x = -3$ and $x = -2$. Since $f(-3) = -27$ which is negative and $f(-2) = 1$ is positive, the graph must cross the x-axis at least once. The function is continuous; thus the curve must connect the two points. To do this, the curve must cross from negative to the positive side of the x-axis. By the definition of continuity, in order for the curve to traverse the axis it must intersect the axis. Each intersection point is called a zero or a root of the function.

Note that the curve must intersect the *x*-axis an odd number of times if it is to pass from the negative side to the positive, for if it traversed the axis an even number of times it would end up on the side on which it started.

Similarly, there is an odd number of roots between $x = 1$ and $x = 2$, and between $x = 3$ and $x = 4$. Furthermore, since the equation is of degree 3, it has exactly three roots. Observe that the curve crosses the *x*-axis three times, indicating the three roots of the equation. Therefore, exactly one root lies in each of the above intervals.

 Find all rational roots of the equation:

$$x^4 - 4x^3 + x^2 - 5x + 4 = 0$$

This is a fourth degree equation. We can solve it by synthetic division. Guess at a root by trying to find an *x*-value which will make the equation equal to zero. $x = 4$ works.

Now write the coefficients of the equation in descending powers of *x*. Note that if a term is missing, its coefficient is zero. In the corner box, the root 4 is placed. Bring the first coefficient down and multiply it by the root. Place the result below the next coefficient and add. Multiply the result by the root and continue as before.

$$
\begin{array}{rrrrr|r}
1 & -4 & +1 & -5 & +4 & \underline{4} \\
  & +4 & 0 & +4 & -4 &   \\
\hline
1 & 0 & 1 & -1 & 0 &   \\
\end{array}
$$

The last result is zero which indicates $(x - 4)$ is a factor and $x = 4$ is a root. The other results are the coefficients of the third degree expression when $(x - 4)$ is factored.

$$(x-4)(x^3 + 0x^2 + x - 1) = 0$$
$$(x-4)(x^3 + x - 1) = 0$$

To find the roots of the third degree equation, call it $g(x)$, we must set it equal to zero.

$$g(x) = x^3 + 0x^2 + x - 1 = 0$$

Try to find where the curve of the equation crosses the x-axis, which is when $y = 0$.

x	-2	-1	0	1	2
y	-11	-3	-1	1	9

It crosses the x-axis between $x = 0$ and $x = 1$. It is an irrational root.

Since the given equation is a fourth degree equation, it has 4 roots. All of the real roots, namely the rational number 4, and an irrational number between 0 and 1, have been found. Therefore, the two remaining roots are not real; that is, they are complex numbers.

 Find the rational roots, if they exist, for the following equation:

$$x^4 - 4x^2 + 5x - 2 = 0.$$

 In general, if an equation $f(x) = 0$ has integral coefficients and is in the form

$$x^n + P_1 x^{n-1} + P_2 x^{n-2} + \ldots + P_{n-1} x + P_n = 0$$

then any rational root of $f(x) = 0$ is an integer and a factor of P_n.

Therefore, the possible rational roots for the given equation are ± 1 and ± 2. To determine whether any of them are the root of the equation, substitute each of these possible roots into the given equation, or use synthetic division. For example, substitute -1 into the given equation.

$$(-1)^4 - 4(-1)^2 + 5(-1) - 2$$

$$= 1 - 4 - 5 - 2 = -10 \neq 0$$

Hence, −1 is not a root of the equation. Similarly, one can show that only 1 is the root of the given equation.

Show that −1 is a root of $f(x) = x^4 + 3x^3 + 4x + 6$.
Hint: Use factor theorem.

The factor theorem says that a polynomial $f(x)$ has an exact factor $x - r$ if and only if $f(r) = 0$.

$$
\begin{array}{r}
x^3 + 2x^2 - 2x + 6 \\
x+1{\overline{\smash{\big)}\,x^4 + 3x^3 + 4x + 6}} \\
\underline{x^4 + x^3} \\
2x^3 \\
\underline{2x^3 + 2x^2} \\
-2x^2 + 4x \\
\underline{-2x^2 - 2x} \\
6x + 6 \\
\underline{6x + 6} \\
0
\end{array}
$$

remainder $= f(r) = f(-1) = 0$

Given that one root of $2x^3 + 4x^2 - 46x - 120 = 0$ is 5, find 2 more roots.

By synthetic division,

Coefficients

or divide

of equation →

$$
\begin{array}{r|rrrr}
& 2 & 4 & -46 & -120 & 5 \\
& & 10 & 70 & 120 & \\
\hline
& 2 & 14 & 24 & 0 & 5
\end{array}
$$

, $\dfrac{\left(2x^3 + 4x^2 - 46x - 120\right)}{(x - 5)}$

Therefore, $2x^2 + 14x + 24 = 0$ is the depressed equation.

Solve the depressed equation

$(2x+6)(x+4) = 0$
$2x+6 = 0 \quad x+4 = 0$
$x = -3 \qquad x = -4$

The roots of the equation are $-3, -4$.

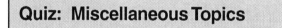

Quiz: Miscellaneous Topics

1. A committee consists of 10 members. How many ways can a subcommittee of 4 members be selected?

   (A) 210.

   (B) 5040.

   (C) 40.

   (D) 256.

   (E) 2060.

2. Expand $(4x - 3y)^5$ using the Binomial Theorem:

   (A) $20x^5 - 15y^5$.

   (B) $1024x^5 + 3840x^4y + 5760x^3y^2 + 4320x^2y^3$
        $+ 1620xy^4 + 243y^5$.

   (C) $1024x^5 + xy - 243y^5$.

   (D) $1024x^5 - 3840x^4y + 5760x^3y^2 - 4320x^2y^3$
        $+ 1620xy^4 - 243y^5$.

   (E) $4x^5 - 60x^4y + 120x^3y^2 - 120x^2y^3 + 60xy^4 - 3y^5$.

3. Out of a 52-card deck, what is the probability of drawing one of the 4 kings?

    (A) $\dfrac{1}{12}$.

    (B) $\dfrac{1}{4}$.

    (C) $\dfrac{4}{1}$.

    (D) $\dfrac{52}{4}$.

    (E) $\dfrac{1}{13}$.

4. Find the value of $\displaystyle\sum_{k=1}^{4} \dfrac{(k+1)}{k(k+2)}$

    (A) $\dfrac{7}{25}$.

    (B) $\dfrac{5}{24}$.

    (C) $\dfrac{91}{60}$.

    (D) $\dfrac{5}{3}$.

    (E) $\dfrac{1}{72}$.

5. What is the 7th term in the sequence $\dfrac{3}{2}, \dfrac{5}{6}, \dfrac{7}{12}, \dfrac{11}{20}, \ldots$

    (A) $\dfrac{13}{42}$.

    (B) $\dfrac{19}{56}$.

    (C) $\dfrac{16}{55}$.

    (D) $\dfrac{17}{54}$.

    (E) cannot be determined.

6. What is the formula that will determine the value of $\dfrac{1}{2} + \dfrac{1}{6} + \dfrac{1}{12} + \ldots + \dfrac{1}{n(n+1)}$ where n is a positive integer?

(A) $\dfrac{n}{n+1}$.

(D) $\dfrac{n^2}{(n+1)^2}$.

(B) $\dfrac{1}{n(n+1)}$.

(E) $\dfrac{3}{(n-1)}$.

(C) $n^2(n+1)$.

7. Find all roots of $f(x) = x^5 + 3x^4 - x - 3$

   (A) $\pm i, \pm 1, -3$ (D) $1, -3$

   (B) $i, 1, 3$ (E) $-1, 3$

   (C) $\pm 1, -3$

8. It is known that $x = \dfrac{2}{3}$ is a zero of $f(x) = 3x^3 - 8x^2 - 5x + 6$. What are the other zeros of $f(x)$?

   (A) $x = -9, x = 3$.   (D) $x = \dfrac{1}{2}, x = \dfrac{-1}{2}$.

   (B) $x = \dfrac{3}{2}, x = -3$.   (E) $x = 3, x = -1$.

   (C) $x = 3, x = -1$.

9.   Which of the following polynomials has zeros $1 \pm i$, 2, -3?

(A) $f(x) = x^4 - x^3 - 8x^2 + 12x$.

(B) $f(x) = x^4 - 3x^3 - 2x^2 + 10x - 12$.

(C) $f(x) = x^4 - x^3 - 6x^2 + 14x - 12$.

(D) $f(x) = x^4 + x^3 - 4x^2 + 7x + 12$.

(E)   None of these.

10.   For the following sequence of numbers, $\dfrac{1}{2}, \dfrac{1}{12}, \dfrac{1}{30}, \ldots,$ the next number will be

(A) $\dfrac{1}{36}$.   (D) $\dfrac{1}{56}$.

(B) $\dfrac{1}{27}$.   (E) $\dfrac{1}{72}$.

(C) $\dfrac{1}{48}$.

ANSWER KEY

1.	(A)	6.	(A)
2.	(D)	7.	(A)
3.	(E)	8.	(E)
4.	(C)	9.	(C)
5.	(B)	10.	(D)

MAXnotes®

REA's Literature Study Guides

MAXnotes® are student-friendly. They offer a fresh look at masterpieces of literature, presented in a lively and interesting fashion. **MAXnotes®** offer the essentials of what you should know about the work, including outlines, explanations and discussions of the plot, character lists, analyses, and historical context. **MAXnotes®** are designed to help you think independently about literary works by raising various issues and thought-provoking ideas and questions. Written by literary experts who currently teach the subject, **MAXnotes®** enhance your understanding and enjoyment of the work.

Available **MAXnotes®** include the following:

Absalom, Absalom!
The Aeneid of Virgil
Animal Farm
Antony and Cleopatra
As I Lay Dying
As You Like It
The Autobiography of
    Malcolm X
The Awakening
Beloved
Beowulf
Billy Budd
The Bluest Eye, A Novel
Brave New World
The Canterbury Tales
The Catcher in the Rye
The Color Purple
The Crucible
Death in Venice
Death of a Salesman
Dickens Dictionary
The Divine Comedy I: Inferno
Dubliners
The Edible Woman
Emma
Euripides' Medea & Electra
Frankenstein
Gone with the Wind
The Grapes of Wrath
Great Expectations
The Great Gatsby
Gulliver's Travels
Handmaid's Tale
Hamlet
Hard Times
Heart of Darkness

Henry IV, Part I
Henry V
The House on Mango Street
Huckleberry Finn
I Know Why the Caged
    Bird Sings
The Iliad
Invisible Man
Jane Eyre
Jazz
The Joy Luck Club
Jude the Obscure
Julius Caesar
King Lear
Leaves of Grass
Les Misérables
Lord of the Flies
Macbeth
The Merchant of Venice
Metamorphoses of Ovid
Metamorphosis
Middlemarch
A Midsummer Night's Dream
Moby-Dick
Moll Flanders
Mrs. Dalloway
Much Ado About Nothing
Mules and Men
My Antonia
Native Son
1984
The Odyssey
Oedipus Trilogy
Of Mice and Men
On the Road

Othello
Paradise
Paradise Lost
A Passage to India
Plato's Republic
Portrait of a Lady
A Portrait of the Artist
    as a Young Man
Pride and Prejudice
A Raisin in the Sun
Richard II
Romeo and Juliet
The Scarlet Letter
Sir Gawain and the
    Green Knight
Slaughterhouse-Five
Song of Solomon
The Sound and the Fury
The Stranger
Sula
The Sun Also Rises
A Tale of Two Cities
The Taming of the Shrew
Tar Baby
The Tempest
Tess of the D'Urbervilles
Their Eyes Were Watching God
Things Fall Apart
To Kill a Mockingbird
To the Lighthouse
Twelfth Night
Uncle Tom's Cabin
Waiting for Godot
Wuthering Heights
Guide to Literary Terms

RESEARCH & EDUCATION ASSOCIATION
61 Ethel Road W. • Piscataway, New Jersey 08854
Phone: (732) 819-8880   **website: www.rea.com**

Please send me more information about MAXnotes®.

Name _____

Address _____

City _____ State _____ Zip _____

REA's Test Preps
The Best in Test Preparation

- REA "Test Preps" are **far more** comprehensive than any other test preparation series
- Each book contains up to **eight** full-length practice tests based on the most recent exams
- **Every** type of question likely to be given on the exams is included
- Answers are accompanied by **full** and **detailed** explanations

REA publishes over 60 Test Preparation volumes in several series. They include:

Advanced Placement Exams(APs)
Biology
Calculus AB & Calculus BC
Chemistry
Computer Science
Economics
English Language & Composition
English Literature & Composition
European History
Government & Politics
Physics B & C
Psychology
Spanish Language
Statistics
United States History

College-Level Examination Program (CLEP)
Analyzing and Interpreting Literature
College Algebra
Freshman College Composition
General Examinations
General Examinations Review
History of the United States I
History of the United States II
Human Growth and Development
Introductory Sociology
Principles of Marketing
Spanish

SAT II: Subject Tests
Biology E/M
Chemistry
English Language Proficiency Test
French
German

SAT II: Subject Tests (cont'd)
Literature
Mathematics Level IC, IIC
Physics
Spanish
United States History
Writing

Graduate Record Exams (GREs)
Biology
Chemistry
Computer Science
General
Literature in English
Mathematics
Physics
Psychology

ACT - ACT Assessment

ASVAB - Armed Services Vocational Aptitude Battery

CBEST - California Basic Educational Skills Test

CDL - Commercial Driver License Exam

CLAST - College Level Academic Skills Test

COOP & HSPT - Catholic High School Admission Tests

ELM - California State University Entry Level Mathematics Exam

FE (EIT) - Fundamentals of Engineering Exams - For both AM & PM Exams

FTCE - Florida Teacher Certification Exam

GED - High School Equivalency Diploma Exam (U.S. & Canadian editions)

GMAT CAT - Graduate Management Admission Test

LSAT - Law School Admission Test

MAT - Miller Analogies Test

MCAT - Medical College Admission Test

MTEL - Massachusetts Tests for Educator Licensure

MSAT - Multiple Subjects Assessment for Teachers

NJ HSPA - New Jersey High School Proficiency Assessment

NYSTCE: LAST & ATS-W - New York State Teacher Certification

PLT - Principles of Learning & Teaching Tests

PPST - Pre-Professional Skills Tests

PSAT - Preliminary Scholastic Assessment Test

SAT

TExES - Texas Examinations of Educator Standards

THEA - Texas Higher Education Assessment

TOEFL - Test of English as a Foreign Language

TOEIC - Test of English for International Communication

USMLE Steps 1,2,3 - U.S. Medical Licensing Exams

U.S. Postal Exams 460 & 470

RESEARCH & EDUCATION ASSOCIATION
61 Ethel Road W. • Piscataway, New Jersey 08854
Phone: (732) 819-8880   **website: www.rea.com**

Please send me more information about your Test Prep books

Name _____

Address _____

City _____ State _____ Zip _____